Commercial Space Capabilities and Market Overview

The Relationship Between Commercial Space Developments and the U.S. Department of Defense

EMMI YONEKURA, BRIAN DOLAN, MOON KIM, KRISTA ROMITA GROCHOLSKI, RAZA KHAN, YOOL KIM

Prepared for the Department of the Air Force
Approved for public release; distribution unlimited

RAND PROJECT AIR FORCE

For more information on this publication, visit **www.rand.org/t/RRA578-2**.

About RAND

The RAND Corporation is a research organization that develops solutions to public policy challenges to h
throughout the world safer and more secure, healthier and more prosperous. RAND is nonprofit, nonpartisa
public interest. To learn more about RAND, visit www.rand.org.

Research Integrity

Our mission to help improve policy and decisionmaking through research and analysis is enabled through o
and objectivity and our unwavering commitment to the highest level of integrity and ethical behavior. To h
and analysis are rigorous, objective, and nonpartisan, we subject our research publications to a robust and ex
process; avoid both the appearance and reality of financial and other conflicts of interest through staff tra
and a policy of mandatory disclosure; and pursue transparency in our research engagements through our c
publication of our research findings and recommendations, disclosure of the source of funding of published
ensure intellectual independence. For more information, visit www.rand.org/about/principles.

RAND's publications do not necessarily reflect the opinions of its research clients and sponsors.

Published by the RAND Corporation, Santa Monica, Calif.
© 2022 RAND Corporation
RAND® is a registered trademark.

Library of Congress Cataloging-in-Publication Data is available for this publication.
ISBN: 978-1-9774-0920-1

Cover: NASA.

About This Report

The U.S. Space Force (USSF) and U.S. Department of Defense (DoD) are examining and pursuing various ways to leverage commercial space capabilities as part of their policy goals to promote the U.S. space industry and their strategy for improving the national security space architecture. As the commercial space industry continues to grow in capability, capacity, and diversity, opportunities for the USSF and DoD to leverage commercial capabilities are expanding. Specifically, the USSF is considering the role of the commercial space industry in its future space architecture and the innovation ecosystem. It is faced with many choices about which commercial capability to leverage or for which military application it should use commercial instead of organic space capabilities.

Making such choices requires a thorough assessment of commercial space capabilities to understand the benefits, risks, and costs associated with using them and inform decisions about trade-offs across those dimensions based on the priorities of the USSF and other relevant stakeholders. To inform decisions related to leveraging commercial space capabilities, the RAND Corporation developed an analytic framework for a systematic and holistic assessment of the benefits, risks, and costs associated with commercial space options.

The results of this research are reported in a series of reports and a spreadsheet tool.

- *Leveraging Commercial Space Capabilities to Enhance the Space Architecture of the U.S. Department of Defense*. This report is not available to the general public.

- *Commercial Space Capabilities and Market Overview: The Relationship Between Commercial Space Developments and the U.S. Department of Defense* (RR-A578-2) characterizes capabilities and trends in the commercial space sector. The cutoff date for information gathered and included in this report is June 15, 2020, and there are rapid changes in the commercial space industry.

- *A Framework for an Integrated Assessment of Commercial Space Capabilities*. This report is not available to the general public.

- *RAND Corporation Spreadsheet Tool to Assess Commercial Capabilities in Space (STACCS)* (TL-A578-1). This tool is not available to the general public.

The research reported here was commissioned by the Office of U.S. Space Force Strategic Requirements, Architectures and Analysis (USSF S5/9)[1] and conducted within the Force

[1] When this project began, the sponsoring office was Air Force Space Command HQ A5/9, which became U.S. Space Force Strategic Requirements, Architectures and Analysis (USSF S5/9). Shortly after the completion of our research, USSF S5/9 was disbanded, with its roles and responsibilities moved to the newly established USSF Headquarters at the Pentagon.

Modernization and Employment Program of RAND Project AIR FORCE as part of a fiscal year 2020 project, *A Robust Strategy for Leveraging Commercial Space Capabilities*.

RAND Project AIR FORCE

RAND Project AIR FORCE (PAF), a division of the RAND Corporation, is the Department of the Air Force's (DAF's) federally funded research and development center for studies and analyses, supporting both the United States Air Force and the United States Space Force. PAF provides the DAF with independent analyses of policy alternatives affecting the development, employment, combat readiness, and support of current and future air, space, and cyber forces. Research is conducted in four programs: Strategy and Doctrine; Force Modernization and Employment; Resource Management; and Workforce, Development, and Health. The research reported here was prepared under contract FA7014-16-D-1000.

Additional information about PAF is available on our website:
www.rand.org/paf

This report documents work originally shared with the DAF on February 3, 2020. The draft report, issued on September 30, 2020, was reviewed by formal peer reviewers and DAF subject-matter experts.

Acknowledgments

We would like to thank Lt Gen William Liquori, Jr., Deputy Chief of Space Operations, Strategy, Plans, Programs, Requirements, and Analysis, Headquarters, USSF, for supporting this project and providing very helpful guidance. We are grateful to our project monitor, Chris Ayres, Space Operations Command, Deputy Commanding General Operations, Deputy Director (formerly USSF S5/9 Technical Director and Advanced Capabilities), for his support, insight, and assistance throughout the project.

We would also like to thank the industry representatives from Intelsat, SES, Amazon Web Services Ground Station, Atlas Space Operations, ExoAnalytic Solutions, Numerica, and Rincon, who participated in our data-collection effort and discussions.

This work also benefited greatly from discussions with Lt Col Paul Muller, a RAND Air Force fellow, about the SATCOM mission, and with RAND colleague George Nacouzi about the remote sensing and space domain awareness missions. We also thank Barbara Bicksler for her communications support to improve the clarity and readability of the report. The content and recommendations of this report, however, are the responsibility of the authors.

Summary

Issue

The U.S. Space Force (USSF) and U.S. Department of Defense (DoD) are considering various ways to leverage commercial space capabilities as part of their policy goals to promote the U.S. space industry and improve the resiliency of the national security space architecture. The commercial space sector offers a range of capabilities and services, including emerging technologies. The commercial capability, commercial capacity, and demand signals from the U.S. government have rapidly evolved recently for many space capabilities. As the USSF and DoD face choices about leveraging commercial space capabilities, they need to be aware of the current capabilities and trends of the commercial space sector.

Approach

The analysis was conducted by reviewing past assessments of commercial space capabilities and open-source literature to characterize capabilities, technological innovation, and trends of the space sector. Where needed, we collected additional new information from relevant government organizations and space service providers. We reviewed the commercial space capabilities of satellite communications (SATCOM), space launch, remote sensing, environmental monitoring, space domain awareness, data transmit/receive networks, and space logistics. Commercial capabilities were limited to U.S. companies or companies with a U.S. subsidiary.

Observations

- **More-established commercial space sectors are growing in capacity and capability**. The SATCOM sector has begun using high-throughput satellites and is planning proliferated low earth orbit constellations. The space launch sector has had two new National Security Space Launch (NSSL)–class entrants, is developing super-heavy launch vehicles (LVs), and has a growing number of small LV entrants. The remote sensing sector has a quickly growing number of multi-satellite constellations and diversity in sensor phenomenology and analytic products.
- New entrants are also responsible for recent growth in the commercial space industry. **The growth and evolution of new entrants have been driven by small satellite technologies and the proliferated constellation model, advanced manufacturing, use of artificial intelligence and machine learning, and venture capital investments**.
- Among the new space sectors, some will serve commercial space operators, while **other new space sectors will primarily target government customers**. Driven by the commercial proliferation of space, space domain awareness entrants will offer enhanced collision warnings, and ground station entrants will provide data-transport services to

offer timely downlink of high-volume data (i.e., for remote sensing satellites). Environmental monitoring entrants are collecting Global Navigation Satellite System Radio Occultation (GNSS-RO) data for the National Oceanic and Atmospheric Agency (NOAA) and DoD. Space logistics entrants are planning space debris–removal services and on-orbit servicing for satellite life extension, both of which garner interest from government space programs.

Recommendations

- All space sectors we reviewed have experienced changes in the past five years (Table S.1), indicating that **it is important for DoD and other stakeholders to periodically update their information about the industry**. These industries changed quite a bit during our research alone, so it may be necessary to update information annually when startups are involved.
- **DoD and other stakeholders should track several technology-development and commercial-viability factors going forward**, because these will have significant impacts on the space market (see Table S.1).

Table S.1. Recent and Future Developments in the Commercial Space Industry

Sector	Changes in Recent Years	Futures to Watch
Satellite communication	• Increased commercial capacity with increased market demand	• Added global broadband capacity from non–geosynchronous satellite operator constellations
Space launch	• Increase in the number of launch-service providers across all launch classes	• Technology developments: reusability, on-orbit reignition, increase lift capacity • Effect of NSSL Phase 2 contract award on the market
Remote sensing	• Expansion in current and planned proliferated low earth orbit launches	• Size of commercial market and financial viability of startups
Environmental monitoring	• NOAA and DoD focus on GNSS-RO • Success in some GNSS-RO launch and operations—commercial and government • Lack of progress in hyperspectral soundings	• New startups with developments in microwave, electro-optical/infrared, and space weather capabilities
Space domain awareness	• Increased demand with more entrants into space domain	• Size of commercial market and financial viability of startups • Space proliferation driving demand and/or collaboration
Data transmit/ receive networks (ground stations)	• New U.S. companies offering ground stations as a service for commercial and government customers	• Electronically steered antennas/multiphase array • Optical communications technology
Space logistics (on-orbit servicing)	• Launch of only one company • Developing niche capabilities from a few companies	• Realization of technological developments, enabling on-orbit refueling, assembly, and manufacturing

SOURCE: RAND analysis of open-source reporting.

Contents

Figures and Tables

Figures

Tables

1. Introduction

The U.S. Space Force (USSF) and U.S. Department of Defense (DoD) are considering and pursuing various ways to leverage commercial space capabilities as part of their policy goals to promote the U.S. space industry and improve the resiliency of the national security space architecture.[1] Currently, the commercial space sector offers a range of capabilities and services that could meet the USSF and DoD's space needs. For example, DoD purchases commercial satellite communications (SATCOM) services and commercial imagery to support a wide range of military applications. Additional opportunities exist with new commercial entrants in other space sectors, such as space domain awareness (SDA), weather, and remote sensing, which are developing innovative capabilities with proliferated constellations of small satellites. Other new sectors, such as on-orbit satellite servicing and ground station data transport, are emerging.

As the commercial space sector is expanding the types of service it offers and is increasing its capabilities with advanced technologies, the USSF and DoD face many choices about which commercial capability to leverage, for which military application they should use commercial capabilities, and how they should acquire those capabilities. Given the importance of these choices, the USSF S5/9 asked RAND Project AIR FORCE to develop a framework for evaluating the opportunities and risks of leveraging commercial space capabilities and make recommendations to help the USSF develop a robust strategy, including risk mitigation, for leveraging commercial space capabilities during all phases of a warfighting conflict. As part of that effort, we conducted a survey of the commercial space sector to characterize capabilities and trends—the subject of this report.

For our analysis, we reviewed mostly past work and open-source literature to characterize capabilities, technological innovation, and trends of the space sector. Where needed, we collected additional new information from relevant government organizations and space service providers.

A previous RAND Corporation report documented commercial space capabilities with data collected through 2016.[2] The authors of that report examined the SATCOM, remote sensing,

---

[1] One of the National Space Policy goals is to promote a robust commercial U.S. space industry. To that end, the policy directs U.S. departments and agencies to purchase and use commercial space capabilities and services to the maximum extent practical. Furthermore, the National Security Space Strategy includes partnering with commercial space entities to improve the resiliency of the national security space architecture. See the Office of the President, *National Space Policy of the United States of America*, Washington, D.C., June 28, 2010; Office of the President, *National Security Strategy of the United States of America*, Washington, D.C., December 2017.

[2] Yool Kim, Ellen Pint, David Galvan, Meagan Smith, Therese Marie Jones, and William Shelton, *How Can DoD Better Leverage Commercial Space Capabilities? Understanding Business Processes and Practices in the Commercial Satellite Service Industry*, Santa Monica, Calif.: RAND Corporation, 2016, Not available to the general public.

SDA (then called *space situational awareness* [SSA]), and on-orbit servicing, with a focus on business models, company planning horizons, and emerging markets beneficial to DoD. That work provided a useful foundation for assessing the available commercial space market. Since 2016, the commercial space capabilities and capacity described in that report have evolved, and new markets have emerged. Our reexamination of the space industry includes these emerging markets, with a close examination of space launch services and data transport ground stations.

In addition to the evolution of the space sector itself, the needs and demand signals from DoD and the USSF for commercial space capabilities have also evolved. Although our research focused on the supply of commercial space capabilities, we acknowledge that the commercial market is influenced by the demand signals it perceives from DoD.

Sponsor guidance informed our review of the following commercial space sectors:

- **satellite communications:** global voice and data connectivity
- **space launch:** delivers payloads, such as satellites, into space, which includes launch range operation
- **remote sensing:** provides information, especially imagery, about Earth using various space-based sensors; relevant to intelligence, surveillance, and reconnaissance
- **environmental monitoring:** provides meteorological, oceanographic, and space weather information to support forecasts, alerts, and warnings
- **SDA:** current and predictive knowledge and characterization of space objects and the space environment
- **data transmit/receive networks:** global ground antenna networks transmit and receive data from satellites to maneuver, configure, operate, and sustain satellite operations
- **space logistics:** satellite rendezvous and proximity operations to support space activities (e.g., propellent depots and satellite servicing).[3]

These missions were chosen to broadly represent areas where there are significant developments from the commercial sector that would be of interest to DoD.[4]

In Table 1.1, we document different factors that would motivate DoD to leverage commercial capabilities for each space sector. Many of these will be discussed further in the sector-specific chapters. The categories of DoD motivations that we document are capacity augmentation, gap filler, increased resilience and flexibility, increased responsiveness, innovation or new capability, faster or more frequent technology refresh, and cost savings. The first four categories reflect the increasing desire for and reliance on space capabilities in DoD missions. Innovation or new capability motivation reflects the growing competition among space companies and the recent developments of the commercial space industry. Generally, a perceived advantage for leveraging a commercial capability is to take advantage of its faster and more-frequent technology refresh

---

[3] These definitions are based on Joint Publication 3-14 mission descriptions, except for remote sensing and space logistics, which were formulated for this research to describe available commercial options (see Joint Publication 3-14, *Space Operations*, Washington, D.C.: Joint Chiefs of Staff, April 10, 2018.

[4] There are other space missions, such as position, navigation, and timing or missile warning, that are not reviewed here because they are outside the project scope.

rates. Additionally, DoD might hope for cost savings by leveraging commercial investments and economies of scale. Both the technology refresh rate and cost savings are motivations across all space sectors; however, the generalization may not always hold true and would require further in-depth cost-benefit analysis of courses of action.[5]

In Table 1.1, each space sector has a unique set of motivational factors to consider. These motivations have and will affect the way DoD chooses to leverage commercial space capabilities. In the chapters on each space sector, we note the various ways DoD or the U.S. government is already engaging the commercial space industry. Ultimately, it is a combination of DoD demand and commercial supply that will determine the future pathways for commercial space partnerships with DoD.

---

[5] In this report, we do not specifically address implementation challenges associated with leveraging commercial capabilities in the different sectors. For more on implementation challenges, see the third report of this series, Yool Kim, Mary Lee, George Nacouzi, Brian Dolan, Moon Kim, and Thomas Light, *A Framework for an Integrated Assessment of Commercial Space Capabilities*, Santa Monica, Calif.: RAND Corporation, forthcoming, Not available to the general public.

Table 1.1. DoD Motivation for Leveraging Commercial Space by Mission

Mission	Capacity Augmentation	Gap Filler	Increased Resilience and Flexibility	Increased Responsiveness	Innovation or New Capability	Faster/Frequent Tech Refresh	Cost Savings[a]
SATCOM	Peacetime and surge capacity	Polar/high latitude coverage	• Narrow beams • Proliferated LEO • Multi-orbit/multiband/multi-network architecture		• Mass manufacturing • Use of alternative frequency bands • Efficient use of spectrum • Use of artificial intelligence/machine learning (AI/ML)	X	X
Space launch			• Access to additional space ports • Access to increased number of launch providers	• Reduced time-lines for launch operations • On-demand launch to support reconstitution	• Reusable multi-orbit launches • 3D printing	X	X
Remote sensing	Priority in image collection		• Proliferated constellations	• Increased revisit rate with proliferated constellations	• Multi-intelligence • Persistence • Use of AI/ML	X	X
Environmental monitoring	Additional atmospheric profile data to feed into forecast models	Potential for continuity gap	• Disaggregated instruments • Backup capability			X	X
SDA	Additional SDA collection sensors around the globe		• Access to increased number of sensors	• Access to new geographic areas	• Analytics to solve SDA knowledge areas • Use of AI/ML	X	X
Data transmit/receive	Augment Air Force Satellite Control Network (AFSCN)		• Access to increased number of ground antennas		• Advanced antenna (phased array, optical) • Use of AI/ML • Use of cloud	X	X
Space logistics (on-orbit servicing)					• On-orbit refuel, repair, or maneuver to reconstitute capability or for defensive operations	X	X

SOURCE: RAND analysis of information provided by the sponsor.

NOTES: The Air Force Satellite Control Network is now called the Satellite Control Network after a name change that occurred after the completion of our research in September 2020.

[a] Cost savings from leveraging commercial investments and economies of scale.

4

Organization of This Report

The chapters in this report present an overview of the results of our research in each of the seven space sectors. Each chapter includes a market overview, a description of market and technology trends, and an assessment of key commercial companies in each sector. In these discussions, we highlight cases where DoD has had influence on the commercial market or, inversely, where the commercial sector has influenced DoD's space mission. We close with a summary of the major changes and issues for commercial space capabilities.

2. Satellite Communications

Mission Scope

The SATCOM market dates to the 1960s and is one of the oldest commercial space sectors. With global revenue of $124.4 billion in 2018,[1] the sector offers a variety of services, including satellite TV and radio, wideband (broadband), and narrowband to consumers, private enterprises, and governments around the world. This chapter focuses on the SATCOM market and select U.S. companies that are considered most relevant for the U.S. government and DoD.[2]

Market Overview

The beginning of the commercial SATCOM market can be traced back to the Communications Satellite Act of 1962,[3] which resulted in the creation of a publicly traded company to establish a commercial SATCOM system. Since then, the SATCOM industry has become one of the few space industries that is characterized by a high level of competition with diversified suppliers and customers. The size of the global SATCOM services market was $24.4 billion in 2018. Fixed satellite services are the largest component of the market, followed by mobile satellite services and consumer broadband, as shown in Table 2.1.

Table 2.1. 2018 Commercial SATCOM Services Market Revenue (Excluding Satellite TV and Radio)

Market Component	Global ($ billions)	United States ($ billions)
Consumer broadband	2.4	2.2
Fixed satellite services	17.9	5.2
Mobile satellite services	4.1	0.62
Total	24.4	8.02

SOURCE: Satellite Industry Association, 2019.

Furthermore, the industry does not depend solely on the government to purchase its goods and services. As shown in Figure 2.1, government revenue, which includes the U.S. and other international governments, for four major SATCOM companies accounted for around 21 percent of the total revenue generated in recent years.

---

[1] Satellite Industry Association, *2019 State of the Satellite Industry Report*, Washington, D.C., May 2019.

[2] Not all companies are headquartered in the United States. Some mentioned in this section are U.S. subsidiaries of foreign parent companies.

[3] Public Law 87-624, Communications Satellite Act of 1962, August 31, 1962.

Nevertheless, the SATCOM industry is an important partner for DoD. In 2019, the department's SATCOM requirement was 30 gigabits per second (Gbps), of which 13 Gbps was supported by the commercial industry.[4] It is estimated that DoD spends about $1 billion per year for SATCOM capacity leases.[5]

Figure 2.1. Revenues of Four Major SATCOM Service Providers

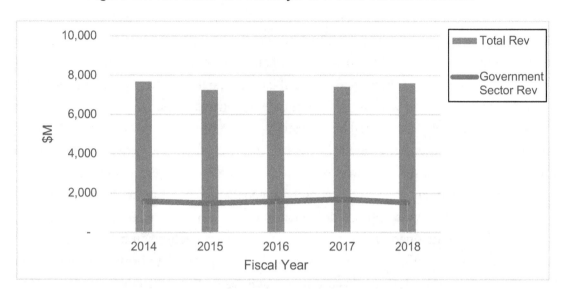

SOURCES: Viasat, *Annual Report 2015*, Carlsbad, Calif., September 2015; Viasat, *Annual Report 2017*, Carlsbad, Calif., September 2017; Viasat, *Annual Report 2019*, Carlsbad, Calif., September 2019; Inmarsat, *In Touch: Inmartsat PLC Annual Report and Accounts 2014*, London, March 2015; Inmarsat, *Building Momentum: Inmarsat PLC Annual Report and Accounts 2016*, London, March 2017; Inmarsat, *Enabling Connectivity: Annual Report and Accounts 2018*, London, March 2019; Intelsat, *Annual Report Pursuant to Section 13 or 15(d) of the Securities Exchange Act of 1934 for the Fiscal Year Ended December 31, 2015*, Washington, D.C.: U.S. Securities and Exchange Commission, commission file number 001-35878, May 2, 2016b; Intelsat, *Annual Report Pursuant to Section 13 or 15(d) of the Securities Exchange Act of 1934 for the Fiscal Year Ended December 31, 2017*, Washington, D.C.: U.S. Securities and Exchange Commission, commission file number 001-35878, February 2018; Intelsat, *Annual Report Pursuant to Section 13 or 15(d) of the Securities Exchange Act of 1934 for the Fiscal Year Ended December 31, 2018*, Washington, D.C.: U.S. Securities and Exchange Commission, commission file number 001-35878, February 20, 2019; SES, *FY 2014 Results: Year Ended 31 December 2014*, February 20, 2015 ; SES, *FY 2016 Results: Year Ended 31 December 2016*, London, February 24, 2017; SES, *Consolidated Financial Statements as at and for the Year Ended 31 December 2018 and Independent Auditor's Report*, Betzdorf, Luxembourg, March 1, 2019.

The total capacity available from the major satellite operators currently under contracts with DoD is estimated to be 254 Gbps, growing to 5,600 Gbps by 2022.[6] The services are

---

[4] Timothy A. Bonds, Frank Camm, and Jordan Willcox, *Ensuring Theater Satellite Communications: Capabilities and Costs of Commercial Services*, Santa Monica, Calif.: RAND Corporation, RR-A103-1, forthcoming.

[5] U.S. Government Accountability Office, "Defense: DoD Commercial Satellite Communication Procurements," in *2016 Annual Report: Additional Opportunities to Reduce Fragmentation, Overlap, and Duplication and Achieve Other Financial Benefits*, Washington, D.C., GAO-16-375, April 13, 2016.

[6] Bonds, Camm, and Willcox, forthcoming.

concentrated in the regions where consumer interests are high. Therefore, although the overall SATCOM capacity provided by the industry exceeds the DoD requirement, limitations of service ability do exist in the regions that lack consumer demand, such as the polar regions.

Trends in Market—Technologies and Capabilities

The SATCOM market has experienced significant changes in technologies and market dynamics in the past few years. The two main technological trends are geosynchronous orbit (GEO) high-throughput satellites (HTSs) that provide an unprecedented amount of per-satellite capacity and the development of the proliferated non–geostationary orbit satellite constellations. Both developments are meant to increase overall capacity, data transfer speed, and accessibility. Such efforts are directed to make SATCOM more competitive in the larger telecommunications market, which is currently dominated by terrestrial service providers through fiber. Concurrent with these developments, several satellite operators are also exploring providing services through multi-orbit models. Because the various orbits have advantages and disadvantages when compared with one another, the goal of using both geostationary orbits and non–geostationary orbits is to seamlessly provide optimized connectivity based on the location and needs of users.

Although the satellite operators have been integrating connections in multiple orbits, ground terminal and antenna manufacturers are working with the operators to bring the connections to users on the ground. In contrast to geostationary satellites that provide connections using wide beams to fixed locations, non–geosynchronous satellite operator (NGSO) satellites rotate around the Earth, and each satellite has a small spot beam that constantly moves. To achieve continuous connection, a terminal must be able to steer and switch from one satellite view to the next. Such a terminal has not yet been developed at a cost-effective margin. In addition, developing NGSO constellations and integrating them with geosynchronous satellite operator (GSO) satellites still needs to occur for the future multi-orbit SATCOM market to be commercially viable.

Along with the technological trends, the market is growing. The fixed satellite services (FSS) sector and the mobile satellite services sector revenues increased from $12.2 billion and $2 billion in 2013[7] to $17.9 billion and $4.1 billion in 2018,[8] respectively. Interestingly, the total revenue of the four companies that constituted about 65 percent of the FSS sector—Intelsat, SES, Eutelsat, and Telesat—has declined, whereas the total revenue for the two dominant mobile satellite service providers has increased, as shown in Table 2.2. Viasat, notably, has increased its market share during this period. The company was one of the first to provide services using HTS,

---

[7] Kim et al., 2016.

[8] Satellite Industry Association, 2019.

8

and its revenue increased from $1.1 billion in 2013[9] to $1.6 billion in 2018,[10] closely trailing Eutelsat as the fourth-largest provider of SATCOM.

Table 2.2. Total Revenue for the Major SATCOM Companies

Company	2013 Revenue ($ billions)	2018 Revenue ($ billions)
Intelsat	2.6	2.2
SES	2.5	2.4
Eutelsat	1.8	1.7
Telesat	0.87	0.7
Inmarsat	1.38	1.47
Iridium	0.38	0.52
Viasat	1.1	1.6

SOURCES: Kim et al., 2016; Intelsat, 2019; SES, 2019; Eutelsat Communications Group, *Consolidated Financial Statements as of 30 June 2019*, Paris, undated; Telesat, *Annual Report Pursuant to Section 13 or 15(d) of the Securities Exchange Act of 1934 for the Fiscal Year Ended December 31, 2018*, Washington, D.C.: U.S. Securities and Exchange Commission, commission file number 333-159793-01, March 1, 2019; Inmarsat, 2019; Iridium Communications, *2018 Annual Report*, McLean, Va., 2019.

Although growing, the SATCOM sector—and its dynamics—continue to change. The line between the FSS and mobile satellite service sectors began to blur in the mid-2010s, when some companies in each sector began providing both types of services.[11] This trend has intensified in the past few years, with Inmarsat offering wideband FSS through its latest Global Xpress fleet, and the traditional FSS providers offering mobile services, such as in-flight connectivity on commercial airlines. The trend is driven by the increase in demand for mobile broadband and the development of very small aperture terminals that created opportunities for FSS and mobile satellite service providers to cross compete.[12]

Despite the increase in the market size, the manufacturing of GEO satellites slowed down. In 2013 and 2014, global commercial orders for GEO satellites were 23 and 21, respectively. The number of orders began to decrease in 2015, down to eight and nine orders in 2017 and 2018, respectively.[13] The decline in the orders was primarily driven by the wait-and-see posture maintained by GEO satellite operators in the wake of NGSO constellations.[14] During this time,

---

[9] Viasat, "ViaSat Announces Record $1.1 Billion in Revenues and $1.4 Billion in Awards for Fiscal 2013," press release, Carlsbad, Calif., May 16, 2013.

[10] Viasat, 2019.

[11] Kim et al., 2016.

[12] Kim et al., 2016.

[13] Satellite Industry Association, 2019.

[14] Jeff Foust, "May the Satellite Industry Live in Interesting Times," *Space Review*, September 17, 2018.

some GEO satellite operators began to enter the NGSO sector. For example, SES completed its purchase of the O3b medium earth orbit (MEO) constellation network in 2016, and Telesat is in the process of developing a low earth orbit (LEO) broadband constellation.

In 2019, the number of commercial GEO satellite orders increased to ten, but the SATCOM sector does not expect to return to the rate of 20-plus orders per year.[15] Instead, a wider variation in the types of satellites is being ordered. Along with the increase in the orders of HTS satellites mentioned previously, demand for reprogrammable, reconfigurable satellites is also growing.[16] Such flexible satellites use digital technologies in channelizers and antennas and allow the operators to reconfigure beam size and service types to be more responsive to the changes in demand.

Lastly, the development of NGSO constellations has brought changes to the manufacturing cycles of the industry. RAND's 2016 study noted that, for FSS satellites, "a typical design life for a communications satellite is 15 years. A typical manufacturing time for a communications satellite is two to three years."[17] Manufacturing individual mobile satellite service satellites takes far less time: Iridium manufactures three satellites per month, and manufacturing a full constellation takes two to three years.[18] The upcoming LEO constellations were planned to take two to three years to manufacture, but individual satellites were planned to be produced at a rate of 15 satellites per week.[19] Furthermore, the estimated life of the LEO satellites is around five years and will require much more frequent upgrades and replacements.

Going Forward

Several indicators should be considered when assessing the future viability of the NGSO market. First, the Federal Communications Commission (FCC) has yet to approve all the satellites that the companies have planned. For example, SpaceX has announced plans to expand its constellation by 30,000 more satellites, and OneWeb has applied for an MEO constellation of 1,280 satellites and an additional 1,980 satellites for its LEO constellation. If the FCC does not approve these additional satellites, the companies will have to reduce the sizes of their constellations.

Second, per FCC regulation, if the NGSO companies are not able to deploy 50 percent of the approved constellation by six years from the date of approval and 100 percent by nine years, the

---

[15] Adrienne Harebottle, "The New, Holistic View of Space," *Via Satellite*, March 2020b; Maxime Puteaux, "Satellite Manufacturing Faces Changes, Uncertainty in Coming Years," *Via Satellite*, May 20, 2020.

[16] Caleb Henry, "Geostationary Satellite Orders Bouncing Back," *SpaceNews*, February 21, 2020b; Puteaux, 2020.

[17] Kim et al., 2016, p. 19.

[18] Kim et al., 2016.

[19] OneWeb Satellites, "Revolutionizing the Economics of Space," webpage, undated.

company loses its license to launch and operate the satellites that are not yet in orbit.[20] Because of the proliferated number of satellites that are approved, this regulation will be a challenging factor for the NGSO companies that need to complete the constellations with the planned number of satellites. Finally, manufacturing and launching thousands of satellites is an endeavor that requires heavy capital expenditure upfront. Therefore, the amount of funding and capital these companies can raise will have a significant effect on the future of these NGSOs. An example of this is OneWeb, a company that underwent bankruptcy. We discuss this in the next section.

Key Company Assessments

The SATCOM market has proven its commercial viability in the past few decades, and the number of suppliers in the sector is abundant and diverse. Globally, more than 40 firms own and operate satellites. A secondary market with capacity resellers is active. From this pool of suppliers, we examined five GSOs and four NGSOs that are considered the key companies in the U.S. SATCOM sector.

GSO Providers

Viasat, SES, Intelsat, Inmarsat, and Eutelsat are the five major companies in the GSO market that have contracts with the U.S. government.[21] The companies primarily operate in the C-, Ku-, and Ka-bands—the frequency group appropriate for high data-rate communication.[22] Although four of the five companies are headquartered in Europe, as shown in Table 2.3, all have subsidiaries in the United States. Some have separate boards for the U.S. government business sector that have a certain level of autonomy from their European parent companies. Collectively, the companies provide services globally, except for the polar regions. Most of their capacity is concentrated in North America, South America, Europe, the Middle East, and Africa,[23] providing services to fixed users and, increasingly, mobile users, including in-flight and maritime connectivity.

Recent developments in satellite technology have significantly increased the capacities of the geostationary satellites. All the companies have begun operating new generations of HTSs that far exceed the previous generations' sub–5 Gbps capacity limit. For example, Intelsat's new

---

[20] Code of Federal Regulations 47, Section 25.161, Automatic Termination of Station Authorization, October 1, 2014.

[21] Eutelsat America Corp and Viasat are contractors to GSA IT Schedule 70. Intelsat General Communications; Inmarsat Government; and SES Government Solutions are contractors to GSA Complex Commercial SATCOM Solutions.

[22] Inmarsat has primarily operated in the narrowband market through L-band. However, in recent years, the company has expanded to operate in the wideband market using Ka-band to serve the increased bandwidth and data rate demand by mobile customers.

[23] Bonds, Camm, and Willcox, forthcoming.

generation of satellites are expected to provide 25 to 60 Gbps per satellite,[24] and Eutelsat's latest Konnect satellite will offer 75 Gbps of capacity.[25] Of the satellites currently in orbit, Viasat-2, launched in 2017, has the highest throughput at 260 Gbps.[26] The company's next satellite, scheduled for 2021, is expected to have 1 Tbps of capacity.[27]

Table 2.3. Key Geostationary Satellite Operators

	Viasat	SES	Intelsat	Inmarsat	Eutelsat
Location/ U.S. subsidiary	California, USA; listed on NASDAQ	Luxembourg/ Washington, D.C., USA	Luxembourg/ Washington, D.C., USA; listed on the New York Stock Exchange	United Kingdom/ Washington, D.C., USA	France/ Washington, D.C., USA
Operational satellites	GEO: 4	MEO: 20 GEO: 57	GEO: 54	GEO: 14	GEO: 37
Frequency band, service quality, service area	C-, Ku-, and Ka-bands 1.5 ~100 Mbps North America, Central America, South America (partial), and the Atlantic maritime routes from North America to Europe	C-, Ku-, and Ka-bands MEO: 500Mbps GEO: Up to 25 Mbps in North America and 15 Mbps elsewhere North America, Latin America, Africa (partial), Europe, Middle East, and Asia (partial), Atlantic Ocean region	C-, Ku-, and Ka-bands Up to 18 Mbps Atlantic Ocean, Americas, Indian Ocean, Asia Pacific, and Pacific Ocean regions	L-, C-, Ku-, and Ka-bands Up to 50 Mbps Globally, excluding the polar regions	C-, Ku-, and Ka-bands Up to 50 Mbps Americas, Europe, Africa, Central Asia, Middle East, Russia, Asia Pacific
Technologies in development	Viasat-3 (2022) serving North and South America, Europe, Africa and Asia Pacific, offering 1 Tbps of total capacity at speed of 100+ Mbps	Using both GEO and MEO to provide faster and higher quality connectivity	Intelsat Epic[NG] platform	GX Flex to provide global coverage, including in the Arctic region	Eutelsat Konnect to launch series of HTSs to provide fast broadband with increased capacity
2018 revenue	$1.6 billion	$2.4 billion	$2.2 billion	$1.3 billion	$1.7 billion
2018 revenue from U.S. and foreign governments	$772 million	$288 million (estimate)	$392 million	$381 million	$187 million

SOURCES: Viasat, "Satellite Fleet," webpage, undated-b; SES, "Explore the Full Fleet," webpage, undated; Intelsat, "Intelsat Coverage Maps," webpage, undated-c; Inmarsat, "Fleet Data," webpage, undated-a; Eutelsat, "Satellites," webpage, undated; Viasat, 2019; SES, 2019; Intelsat, 2019; Inmarsat, 2019; Eutelsat, 2019.

---

[24] Intelsat, *Intelsat Epic[NG]*, September 2016a.

[25] Eutelsat, "Successful Launch of Eutelsat Konnect," press release, Paris, January 17, 2020.

[26] Caleb Henry, "Dankberg Teases ViaSat-4 Specs, Still Mulling MEO Constellation," *SpaceNews*, October 16, 2019d.

[27] Viasat, "KA-SAT Satellite," webpage, undated-a.

Furthermore, the GSO market is evolving through partnerships and expansions to new markets. In addition to the satellites in the geostationary orbit, SES operates a 20-satellite constellation in MEO. Intelsat was preparing for a partnership with OneWeb, a company developing a proliferated LEO constellation, although the deal was canceled. Lastly, Inmarsat, traditionally a GSO narrowband mobile services provider primarily for maritime vessels, has initiated its wideband mobile service with its new generation HTS constellation and continues to develop it through the 2020s. The company plans to provide global wideband coverage, including to the Artic, which had been an excluded region from the GSO market because of lack of commercial demand.[28]

NGSO Providers

The significant increase in capacity from the GSO market will be supplemented by even greater capacity from the NGSO constellations. The NGSOs are developing various sizes of constellations, ranging from hundreds to a few thousand satellites. Individual satellites will have less capacity compared with GSO HTS, but the proliferated nature of the NGSO satellites allows the operators to provide significant capacity to the market.

Operating proliferated constellations of small satellites in non–geostationary orbits has several advantages compared with operating small constellations of large satellites. For example, because of proximity to the users on earth, satellites in LEO and MEO can reduce the latency of communication and provide faster data rates. As shown in Table 2.4, the NGSOs have successfully demonstrated high-speed broadband connectivity with latency in the range of 20 to 40 milliseconds. Also, because the satellites in the non–geostationary orbits are constantly orbiting the Earth, the NGSO constellations can provide services globally when enough satellites are strategically allocated in various inclinations.

In recent years, numerous companies have sought FCC approvals to develop NGSO communication satellite constellations. Some companies have ceased operations before launching their first satellite, and now fewer companies are in the market. Of these companies, three NGSOs—OneWeb, SpaceX, and Telesat—have made significant progress in 2019. Although these three firms have the same objective of providing global satellite broadband, each has unique competitive advantages.

---

[28] Inmarsat, "Global Xpress," webpage, undated-b.

13

Table 2.4. Key Non–Geostationary Satellite Operators

	OneWeb	SpaceX	Telesat
Location/U.S. subsidiary	Headquarters – United Kingdom U.S. subsidaries – Virginia, California, and Florida	California, USA	Headquarters – Canada U.S. subs – New Jersey and Washington, D.C.
Federal Aviation Administration (FAA)–approved constellation	LEO: 720	LEO: 4,409 Very LEO: 7,518	LEO 1: 117 LEO 2: 117
Frequency band	Ku and Ka	LEO: Ku and Ka Very LEO: Ka and V	LEO 1: Ka LEO 2: V
Service-quality demonstrated	Demonstrated 400 Mbps with 32 ms latency	Demonstrated 610 Mbps of download speed	Demonstrated 370 Mbps of downlink and 110 Mbps of uplink with 20–40 ms latency
Estimated total network capacity of approved constellations	6 Tbps	200 Tbps	6 Tbps
Pending FCC approval	Request to increase to the LEO constellation to 47,844 satellites MEO constellation of 1,280 satellites	None with FCC currently, but applied to International Telecommunication Union for an additional 30,000 satellites	Information not available
Initial service offering	Artic region in late 2020 Global coverage in 2021	Northern United States and Canada in 2020 Global coverage in 2021	2022

SOURCES: OneWeb, "Technology," webpage, undated; Caleb Henry, "Musk Says Starlink 'Economically Viable' with Around 1,000 Satellites," *SpaceNews*, May 15, 2019b; Telesat, "Telesat LEO – Why LEO?" webpage, undated; FCC, "Order and Declaratory Ruling: In the Matter of WorldVu Satellites Limited, Petition for a Declaratory Ruling Granting Access to the U.S. Market for the OneWeb NGSO FSS System," Washington, D.C., FCC-17-77, June 22, 2017a; FCC, "Report and Order and Further Notice of Proposed Rulemaking: In the Matter of Update to Parts 2 and 25 Concerning Non-Geostationary, Fixed-Satellite Service Systems and Related Matters," Washington, D.C., FCC-17-122, September 26, 2017b; FCC, "Order and Declaratory Ruling: In the Matter of Telesat Canada, Petition for Declaratory Ruling to Grant Access to the U.S. Market for Telesat's NGSO Constellation," Washington, D.C., FCC-17-147, November 2, 2017c; FCC, "Memorandum Opinion, Order and Authorization: In the Matter of Space Exploration Holdings, LCC, Application for Approval for Orbital Deployment and Operating Authority for the SpaceX NGSO Satellite System," Washington, D.C., FCC-18-38, March 28, 2018b; FCC, "Memorandum Opinion, Order and Authorization in the Matter of Space Exploration Holdings, LCC, Application for Approval for Orbital Deployment and Operating Authority for the SpaceX V-Band NGSO Satellite System," Washington, D.C., FCC-18-161, November 15, 2018c; FCC, "FCC Application for Space and Earth Station: MOD or AMD," Washington, D.C., File Number SAT–MOD–20180319−00022, March 19, 2018a; FCC, "Order and Authorization in the Matter of Space Exploration Holdings, LCC, Request for Modification of the Authorization for the SpaceX NGSO Satellite System," Washington, D.C., FCC-DA-19-342, April 26, 2019a; FCC, "Application for Modification, in the Matter of WorldVu Satellites Limited, Modification to OneWeb U.S. Market Access Grant for the OneWeb Ku- and Ka-Band System," Washington, D.C., SAT-MPL-20200526-00062, May 26, 2020; Henry, 2019b.

First, OneWeb is a newly founded company that has been approved by the FCC for a 720-satellite constellation in LEO. The company has made strategic partnerships with firms and investors to horizontally integrate the business from manufacturing to distribution of capacity. For example, the company formed a joint venture with Airbus to create a manufacturing unit that plans to build 15 satellites per week.[29] Also, the largest investor of the company holds the rights to sell the capacity from the constellation. OneWeb currently has 74 satellites[30] in orbit and aimed to provide initial service by the end of 2020.[31]

In March 2020, while this project was being conducted, OneWeb filed for bankruptcy because of difficulties in raising additional funding needed to continue business. A buyer is expected to emerge to acquire OneWeb's assets—including the satellites in orbit and the spectrum licenses from the FCC and the International Telecommunication Union—through the bankruptcy proceedings. A buyer could continue the development of the constellation for broadband connection. This major development highlights why it is important for DoD to periodically update its information about the commercial sector.[32]

Whereas OneWeb is horizontally integrated, SpaceX is a vertically integrated company. The launch service provider has been approved by the FCC for two constellations of 11,927 total satellites. The company manufactures its own satellites and uses its competitive advantage as a launch service provider to launch at a lower cost and on its own schedule. SpaceX already has launched 540 satellites[33] and was planning to launch satellites, on average, twice a month in 2020 and begin providing service by the end of the year.[34] In terms of total network capacity, the two SpaceX constellations have a theoretical capacity of 240 Tbps,[35] which is significantly more than all the current GSO satellites combined.

The third company, Telesat, is a GSO company that has been approved by the FCC to operate two 117-satellite constellations in LEO. Unlike the other companies, Telesat does not manufacture its satellites but has extensive experience in operating communication satellites

---

[29] OneWeb Satellites, undated.

[30] As of June 10, 2020.

[31] Darrell Etherington, "Watch OneWeb Launch 34 Satellites for Its Broadband Constellation Live," *Tech Crunch*, February 6, 2020.

[32] During preparation of the final report, we learned that, on October 2, 2020, a federal bankruptcy court approved the sale of OneWeb to the British government and Indian telecommunications company Bharti Global (Rachel Jewett, "U.S. Bankruptcy Court Approves OneWeb Sale to UK Government, Bharti," *Via Satellite*, October 5, 2020).

[33] As of June 18, 2020, and counting two prototypes.

[34] Sandra Erwin, "Starlink's Busy Launch Schedule Is Workable, Says 45th Space Wing," *SpaceNews*, January 7, 2020a.

[35] Debopam Bhattacherjee, Waqar Aqeel, Ilker Nadi Bozkurt, Anthony Aguirre, Balakrishnan Chandrasekaran, P. Brigten Godfrey, Gregory Laughlin, Bruce Maggs, and Ankit Singla, "Gearing Up for the 21st Century Space Race," *HotNets '18: Proceedings of the 17th ACM Workshop on Hot Topics in Networks*, Redmond, Wash., November 15–16, 2018.

since 1969. Although it has not selected the manufacturer of its satellites, the company has demonstrated important technologies, such as the integration of satellites to the terrestrial 5G network,[36] with its two test satellites in orbit. After selecting its manufacturer in 2020, Telesat plans to provide initial broadband service in 2022 and full global service in 2023.[37]

---

[36] Adrienne Harebottle, "Heading into the LEO Revolution," *Via Satellite*, February 2020a.

[37] Chris Forrester, "Telesat LEO Constellation in 2022," *Advanced Television*, March 2, 2020.

3. Space Launch

Mission Scope

The space launch sector provides transportation to space for all other commercial space sectors, making it critical to U.S. military, civil, and commercial space operations. Various policies, including the National Security Strategy and those articulated in the U.S. Code, highlight the importance to the country of assured access to space. Numerous launch market segments exist, and each provides various levels of lift capacities. The two market segments most relevant to the needs of DoD are the medium- to heavy-lift launch segments, which includes the National Security Space (NSS) payloads. The emerging small-lift launch market segment primarily caters to scientific payloads and is included in this review.

Market Overview

The medium- to heavy-lift class and the small-lift class markets have different characteristics and different relationships with the U.S. government, as shown in Table 3.1. The medium- to heavy-lift class is categorized by a wide range of orbits and a large payload capacity. An example of the mass-to-orbit capabilities supported in this market is the 12 different reference orbits with payload mass ranging from 5,000 to 37,500 pounds required by DoD.[1] The launch vehicles (LVs) that have been certified by the U.S. Air Force for this class are SpaceX's Falcon 9 and Falcon Heavy vehicles and United Launch Alliance's (ULA's) Atlas V and Delta 4 vehicles. These LVs can provide services to military, civil, and commercial clients. For the upcoming National Security Space Launch (NSSL) Phase 2 program (NSSL-2), two more launch providers, along with SpaceX and ULA, have submitted their response to the request for proposal FA8811-19-R-00020, which closed in February 2020. The two other launch providers are Northrop Grumman, which already operates a medium LV but not NSSL grade, and Blue Origin, a new entrant to the market.

---

[1] Bonnie L. Triezenberg, Colby Peyton Steiner, Grant Johnson, Jonathan Cham, Eder Sousa, Moon Kim, and Mary Kate Adgie, *Assessing the Impact of U.S. Air Force National Security Space Launch Acquisition Decisions: An Independent Analysis of the Global Heavy Lift Launch Market*, Santa Monica, Calif.: RAND Corporation, RR-4251-AF, April 2020.

Table 3.1. U.S. Space Launch Market Overview

Market	Medium to Heavy Lift	Small Lift
Number of launches by U.S. commercial launch service companies (2006–2018)	205	28
Number of U.S. companies in operation	3 ULA, Northrop Grumman, SpaceX	2 Rocket Lab, Northrop Grumman (formerly Orbital Sciences)
Number of new entrants in development	1 Blue Origin	20+
Main customer base	U.S. government	Small satellite operators

SOURCES: RAND analysis of open-source reporting from NewSpace Index, "Welcome to NewSpace Index," website, undated-b; 30th Space Wing, "Western Range Launch Database 2003–2018," database, September 2019, Not available to the general public; 45th Space Wing, "Eastern Range Launch Database 2003–2018," database, August 2019, Not available to the general public; and FAA, "Licensed Launches," webpage, August 18, 2020.

The U.S. commercial launch industry has been around since the 1990s, and the medium- to heavy-lift market depends heavily on the U.S. government. Of the 226 launches conducted from 2006 to 2018 by the main families of vehicles in this market,[2] 169 were U.S. government payloads, both national security and civil.[3] Although the launch providers in this class serve commercial customers, the major source of demand is the U.S. government—a trend that is likely to continue for this class of LVs. Expected demand is about 14–21 launches per year, of which only four to seven would be for commercial customers in the long term.[4] In terms of capacity, the four launch companies that are bidding for the NSSL-2 contracts are estimated to be planning for the combined capacity of 48–60 launches per year when all the vehicles are in operation.[5]

In contrast, the small-lift sector is driven by commercial demand. With payload capacity of less than 2,000 kg to LEO, the LVs in this class are designed to address the growing demand from the operators of small commercial satellites and proliferated LEO constellations. Globally,

---

[2] Atlas, Delta, Falcon, and Space Shuttle launches were conducted during this period for the NSS launches. Antares is another U.S. LV that operated in this period; it conducted seven launches from 2014 (first flight) to 2018. However, Antares was excluded from the count of the launches in this chapter for consistency, because it is not certified by the Air Force to launch NSS payloads. All the Antares launches have been for NASA payloads. Furthermore, the 205 launches in Table 3.1 represent launches conducted by commercial launch service providers and exclude the 21 shuttle launches that occurred between 2006 and 2018.

[3] 30th Space Wing, 2019; 45th Space Wing, 2019.

[4] Triezenberg et al., 2020.

[5] Gary McLeod, Ellen Pint, Eric Larson, Eder Sousa, and Jonathan Tran, *U.S. Space Launch Locations to Support the National Security Space Launch Program: An Independent Assessment of the Ability of the Eastern and Western Ranges to Support Forecasted Launch Demands*, Santa Monica, Calif.: RAND Corporation, 2019, Not available to the general public. During the preparation of this final report, we learned that the NSSL-2 awards were given to SpaceX and ULA in August 2020 (see Sandra Erwin, "Pentagon Picks SpaceX and ULA to Remain Its Primary Launch Providers," *SpaceNews*, August 7, 2020d.

more than 40 LVs are under development in various stages.[6] These small-lift launch service providers aim to provide more flexible and frequent launch schedules customized for small satellites, which have been launched as rideshares on larger payloads launched by heavier classes of LVs. The Air Force's program of record for using the small-lift launch services sector is the Orbital Services Program-4, which is aimed at smaller and more risk-tolerant payloads.

Trends in Market—Technologies and Capabilities

As mentioned previously, the number of suppliers in the launch market is growing. In the medium- to heavy-lift sector, the United States historically had only one or two launch companies providing launch services. In the 2010s, the market began to change, when SpaceX and Orbital Sciences entered the market. The U.S. medium- to heavy-lift sector is now projected to have four operational launch service providers, listed in Table 3.1. The growth in the suppliers in the small-lift sector is even more significant, with the number growing from one (Orbital Sciences) in the early 2010s to potentially more than 20 in the near future.

In the medium- to heavy-lift sector, the main technological trend is reusability. Three of the four companies for the NSSL-2 program have proposed LVs with reusable first-stage boosters, which makes these systems more cost effective. Therefore, these companies are likely to invest more into the technologies that enable higher rates of reusability. Furthermore, SpaceX is developing a new two-stage launch system with reusable first and second stages. With this new system, the second stage, named Starship, will reenter Earth and land at a place where it can be refueled, refurbished, and relaunched. This improvement in reusability would allow the company to be more cost effective.

In the small-lift sector, the main technology trend is 3D printing with additive manufacturing. Advancements in 3D printing have enabled rapid prototyping and a reduction in the number of parts needed, all leading to cost efficiency. Along with the companies just mentioned, many other small-lift LVs under development are using 3D printing in the manufacturing process. The medium- to heavy-lift LVs also use 3D printing to a certain degree. For example, SpaceX prints engine chambers, and Blue Origin prints oxidizer pumps.[7]

Going Forward

The four companies participating in the NSSL-2 request for proposal may have the capacity and capability to serve both commercial and U.S. government clients, but only two companies will be selected for the NSSL-2 program. The NSSL-2 program will be the USSF's major

---

[6] Satellite Industry Association, *State of the Satellite Industry Report*, Washington, D.C., May 2019.

[7] Raman Ponnappan, "Additive Manufacturing in Launch Vehicles," *Spacetech Asia*, August 30, 2018. Small-lift companies and medium- to heavy-lift companies differ in that the former group's effort is in drastically increasing the use of 3D printing technologies to manufacture the LVs, such as developing an engine with only three parts, compared with printing smaller components.

program of record for space launches from 2022 to 2027. Although the companies that are not selected for NSSL-2 can compete for NASA contracts, they cannot participate in USSF launches. Because the major source of demand for the NSSL-class LVs is the U.S. government, split between the USSF and NASA, not being able to participate in NSSL-2 will create challenges for the future of the remaining companies.[8]

Launch-failure rate is another important indicator of the viability of launch companies. Too many failures are detrimental to product reliability and can force a company to exit the market. Also, because the launch market is becoming more competitive, with more companies entering the sector, customers now have more options for service providers. As the technologies in quality assurance and mission safety are constantly improving, launch-failure rates are decreasing. In turn, any launch failure becomes even more detrimental to a company's business viability.

An indicator specific to the small-lift launch market is the rideshare programs by the larger launch-service providers. Several companies in the intermediate- and heavy-lift sectors have announced plans to provide affordable rideshare programs to small payloads at prices competitive to the small-lift launch services.[9] Also, NASA and the USSF are actively seeking opportunities for small satellite developers to share slots on larger LV launches to help with the development of the small satellite market.[10] Although the small-lift launch-service providers have a competitive advantage in scheduling flexibility compared with the rideshare programs that follow the primary payload schedules, such programs take away from demand that otherwise could be provided by the small-lift market.

Key Company Assessments

NSSL-Class Launch Providers

Four companies—Blue Origin, Northrop Grumman, SpaceX, and ULA—have submitted bids for NSSL-2, the main program of record for USSF space launches from 2022 to 2026.[11] For this upcoming program, SpaceX offered Falcon 9 and Falcon Heavy LVs, both of which have launched military payloads under the previous Air Force NSSL launch contracts. Falcon 9 is a two-stage LV with a reusable first-stage booster with a payload capacity of 22,800 kg to LEO and 8,300 kg to geostationary transfer orbit (GTO).[12] As of March 1, 2020, Falcon 9 was

---

[8] As previously mentioned, the NSSL-2 award decision was made after this report's information cutoff date. The award was given to ULA and SpaceX, leaving Blue Origin and Northrop Grumman without government funding.

[9] Jeff Foust, "Opportunities Grow for Smallsat Rideshare Launches," *SpaceNews*, February 6, 2020c.

[10] Debra Werner, "Government Agencies Prepare for Piggyback Flights, Secondary Payloads," *SpaceNews*, September 17, 2018.

[11] Sandra Erwin, "ULA, SpaceX, Blue Origin, Northrop Grumman Submit Bids for National Security Launch Procurement Contract," *SpaceNews*, August 12, 2019a.

[12] SpaceX, "Falcon 9," webpage, undated-a.

launched by SpaceX 79 times and succeeded 77 times, with the last launch failure in 2016. Falcon Heavy is also a two-stage LV with reusable first stages, which consist of three Falcon 9 first stages. With three successful launches thus far, Falcon Heavy can lift 63,800 kg to LEO and 26,700 kg to GTO.[13]

Both ULA and Northrop Grumman have operational LVs but are developing new generations of LVs for NSSL-2. ULA's Vulcan Centaur, with an expected first launch scheduled for 2021, will be ULA's first reusable first-stage LV.[14] Various configuration designs are currently planned, ranging from 10,600 to 27,200 kg to LEO and 7,600 to 14,400 kg to GTO.[15] The company has had a 100-percent mission success rate with its Atlas and Delta LVs, with more than 135 launches since 2006.[16] ULA also offered Atlas V for NSSL-2 in case its new LV is not ready in time.

Whereas the other companies are focusing on reusability as a means to decrease cost, Northrop Grumman is developing a simplified expendable system for its OmegA LV.[17] OmegA is a three-stage LV designed to lift payloads of up to 10,100 kg to GTO.[18] Blue Origin is developing New Glenn, a two-stage LV with a reusable first stage and a payload capacity of 45,000 kg to LEO and 13,000 kg to GTO.[19] New Glenn is the company's first orbital vehicle and is expected to launch in 2021. The vehicle will be powered by the company's own BE-4 engines, which are also used for ULA's Vulcan LVs.

Launches to support the U.S. government's national security space mission are the most sophisticated and stressing in terms of technical capabilities and payload capacity. The LVs under development for the NSSL also will allow the companies to participate in the commercial medium- to heavy-lift market. However, because USSF is one of the major customers for the U.S. commercial launch market, the outcome of the NSSL-2 selection is likely to influence the future outlook of these launch providers.

Small-Lift Launch Providers

Advancements in satellite and computer technologies have allowed significant developments in smaller satellites in the last decade, resulting in a substantial increase in the number launched

---

[13] SpaceX, "Falcon Heavy," webpage, undated-b.

[14] Caleb Henry, "ULA Gets Vague on Vulcan Upgrade Timeline," *SpaceNews*, November 20, 2019e.

[15] ULA, "Rocket Rundown: A Fleet Overview," technical summary, 2019.

[16] ULA, "About," webpage, undated.

[17] Sandra Erwin, "Northrop Grumman Touts Financial Strength in Marketing Pitch for OmegA Rocket," *SpaceNews*, December 3, 2019c. Northrop Grumman announced on September 9, 2020, that it will not continue development of its OmegA rocket (see Sandra Erwin, "Northrop Grumman to Terminate OmegA Rocket Program," *SpaceNews*, September 9, 2020e).

[18] Justin Davenport, "NGIS OmegA Fires for Two Minutes in First Static Test – Nozzle Incident Under Review," *NASA Space Flight*, May 30, 2019.

[19] Blue Origin, "New Glenn," webpage, undated.

in recent years. In 2018, the number of satellites launched that were less than 500 kg doubled to about 320 satellites from the average of 160 satellites per year from 2013 to 2017.[20] The number is expected to grow to 880 satellites per year by 2028.[21] This increase in the small satellite market has catalyzed the growth in the small-lift launch sector.

Following the increase in demand, the sector is also experiencing a substantial growth in potential suppliers. According to a publicly available database, 98 companies with plans to provide small-lift launch service have been founded globally since 2010, compared with nine companies founded in the previous decade.[22] Of these companies, 71 are in the development stage, six have operational LVs, and the remaining 21 are inactive.[23] About half of these companies (48) are registered in the United States.

Although the trend in the number of small satellites launched is increasing, the ability of the market to support a significant number of suppliers is uncertain. Small satellite launches occur in batches, which means that 880 satellites do not equate to the same number of launches. For example, a small-lift LV that has a payload capacity of 1,000 kg can launch two 500 kg small satellites or 100 nano satellites (less than 10 kg). Furthermore, medium- to heavy-lift launch services offer rideshare programs, with prices comparable to the small-lift vehicles; governments often offer free rides to small satellites developed by universities and labs. This competition may be offset by the fact that small-lift launch services are focused on launch responsiveness.[24] Nonetheless, four U.S. companies, as summarized in Table 3.2, stand out in terms of development stage and business strategy in this potentially congested market.

---

[20] Maxime Puteaux and Alexandre Najjar, "Analysis | Are Smallsats Entering the Maturity Age?" *SpaceNews*, August 6, 2019.

[21] Euroconsult, "Euroconsult Research Projects Smallsat Market to Nearly Quadruple over Next Decade," press release, Paris, Washington, D.C., Montreal, Yokohama, August 5, 2019.

[22] NewSpace Index, "Small Satellite Launchers," webpage, undated-a.

[23] Inactive status includes in concept, dormant, and canceled.

[24] Here, the term *responsiveness* refers to the idea of dedicated launches for small satellite operators versus rideshares on larger vehicles. Dedicated launches allow the small satellite operators more control of schedule compared with rideshares (Jeff Foust, "Small Launch Vehicle Companies See Rideshares as an Opportunity and a Threat," *SpaceNews*, February 7, 2019b).

Table 3.2. Key U.S. Small-Lift Launch Service Providers

	Rocket Lab	Firefly Aerospace	Relativity Space	Virgin Orbit
LV (in development)	Electron	Alpha	Terran 1	LauncherOne
Lift capability	225 kg to SSO	1,000 kg to LEO 630 kg to SSO	1,250 kg to LEO 700 kg to SSO	500 kg to LEO 300 kg to SSO
Specialty	3D-printed engine with goal to manufacture one LV per week	Simpler engine design	3D printing of 95 percent of rockets	Air-launch
Cost per launch (cost per kg to LEO)	$5 m ($22,222 per kg)	$15 m ($15,000 per kg)	$10 m ($8,000 per kg)	$10 m–12 m ($20,000–24,000 per kg)
Initial service offering	2018	2020	2021	2020
Mission success rate	12/12	Not applicable	Not applicable	Not applicable
Government clients served	NASA, Air Force, and Defense Advanced Research Projects Agency (DARPA)	Not applicable	Not applicable	Not applicable
Launch sites	Wallops Flight Facility/Mid-Atlantic Regional Spaceport, New Zealand	Cape Canaveral Air Force Station, Florida; Vandenberg Space Force Base, California	Cape Canaveral Air Force Station, Florida; Vandenberg Space Force Base, California	Mojave Desert, California; Kennedy Space Center, Florida; Guam

SOURCES: RAND analysis of open-source reporting from FAA, 2020; Rocket Lab, "Completed Missions," webpage, undated; Firefly Aerospace, "Firefly Alpha," webpage, undated; Relativity Space, "Terran 1," webpage, undated; Virgin Orbit, "LauncherOne," webpage, undated. All data are accurate as of our review cutoff date of June 15, 2020.

Rocket Lab, founded in 2006, is the leading launch-service provider in this sector. Since its first flight in 2018, the company's active LV, Electron, has completed 12 missions,[25] including payloads from NASA, the Air Force, and DARPA.[26] With the capacity to lift up to 200 kg to SSO, the LV has the advertised price of $7.5 million. The primary launch sites are Wallops Flight Facility in Virginia and Mahia Launch Complex in New Zealand. The company manufactures its engine through 3D-printing technology and plans to scale its production to deliver one launch per week.[27]

Two companies, Firefly Aerospace and Relativity Space, are developing LVs larger than Electron. Firefly is developing an LV, Alpha, that can launch up to 1,000 kg to LEO for $15 million.[28] The company made a series of noticeable business arrangements after experiencing liquidation and a change of ownership in 2017. For example, it was awarded indefinite delivery/indefinite quantity launch contracts from NASA and the Air Force and made long-term

---

[25] As of June 12, 2020.

[26] Rocket Lab, undated.

[27] Adam Mann, "Rocket Lab's Electron Rocket," *Space*, October 3, 2019.

[28] Firefly Aerospace, undated.

lease agreements for launch complexes at Vanderburg Space Force Base and Cape Canaveral. Alpha was scheduled to debut in 2020.[29] Relativity Space's Terran 1, scheduled to begin operation in 2021, will have launch capability of 1,250 kg to LEO at $10 million per launch.[30] The company plans to 3D print its LV and reduce the number of parts significantly. Specifically, through advanced additive manufacturing, Relativity's Aeon engine is being developed to consist of only three parts versus the thousands of parts required to make traditional rocket engines, allowing for faster and cost-efficient production.[31]

The final company, Virgin Orbit, has a different plan for launching small satellites. LauncherOne is designed to launch satellites from Boeing 747 jumbo jets midair to LEO, with a capability of 500 kg to LEO.[32] With initial service expected to begin in 2020, the launch sites are in the Mojave Desert in California, the Kennedy Space Center in Florida, and Guam. However, with the ability to launch from an airplane, the company hopes to provide a wide range of options in the future using a network of airports.[33] The price per launch is expected to be between $10 million to $12 million.

---

[29] Elizabeth Howell, "Firefly Aerospace Preps for Debut Flight of Its Alpha Rocket in April," *Space*, January 6, 2020.

[30] Jeff Foust, "Relativity Space Raises $140 Million," *SpaceNews*, October 1, 2019c; Relativity Space, undated.

[31] Bryce Salmi, "The World's Largest 3D Metal Printer Is Churning Out Rockets," *IEEE Spectrum*, October 25, 2019.

[32] Virgin Orbit, undated.

[33] Jeff Foust, "Virgin Orbit Nearing First Launch," *SpaceNews*, February 5, 2020b.

4. Remote Sensing

Mission Scope

Remote sensing is defined as the acquiring of information from a distance.[1] Because this can be done in many ways, remote sensing satellites can host a variety of sensors that include active sensors—radar and scatterometers—and passive sensors, such as radiometers and spectrometers. For the purposes of this report, the sensors affiliated with remote sensing have been limited to those with terrestrial imaging capabilities, which have the most relevance to the intelligence, surveillance, and reconnaissance (ISR) mission. Such sensors include multispectral imagers, hyperspectral imagers, panchromatic imagers, synthetic aperture radar, and short-wave infrared (SWIR) cameras. This chapter provides an overview of the current status and potential of the remote sensing market and closely examines a small number of key players.

Market Overview

History of Commercial Remote Sensing Market

Limited civil and commercial space-based remote sensing efforts in the 1970s and 1980s generated significant U.S. government interest in supporting the growth of a commercial market.[2] In 1992, the Land Remote Sensing Policy Act established a foundation for building the licensing framework for commercial remote sensing and created an environment for private attempts at creating commercial remote sensing businesses.[3] The act incentivized private firms to enter the market by allowing them to sell images to private consumers at market rates. Although the first set of final National Oceanic and Atmospheric Agency (NOAA) licensing rules would not be published until 2000, several private companies entered the market during the 1990s.

In the early 2000s, the U.S. government took additional steps to support the growth of a commercial remote sensing satellite market. In 2003, the Commercial Remote Sensing Space Policy (also known as National Security Presidential Directive 27)[4] required that NOAA clarify and improve its licensing and regulatory framework and directed government agencies to

---

[1] NASA, "What Is Remote Sensing," NASA Earth Data, April 20, 2020.

[2] Landsat was the first civil U.S. remote sensing program, with initial launch in 1972. In 1984, the Earth Observation Satellite Corporation won a contract to privately operate Landsat, market its imagery products, and develop Landsat 6 and 7. The company's failure to create a successful commercial business from Landsat partially drove the passing of the Land Remote Sensing Policy Act of 1992.

[3] Public Law No. 102-555, Land Remote Sensing Policy Act of 1992, October 28, 1992.

[4] White House, "Fact Sheet: U.S. Commercial Remote Sensing Policy," press release, Washington, D.C.: U.S. Commercial Space Remote Sensing Space Policy, May 13, 2003.

[r]ely to the maximum practical extent on U.S. commercial remote sensing space capabilities for filling imagery and geospatial needs for military, intelligence, foreign policy, homeland security, and civil users.

Also in 2003, the National Geospatial-Intelligence Agency (NGA) announced its ClearView initiative to award multiyear contracts to U.S. commercial imagery providers; three initial contracts were awarded to DigitalGlobe, Space Imaging, and OrbImage.[5] OrbImage changed its name to GeoEye in 2006, when it acquired Space Imaging. In 2012, budget cuts prompted NGA to cancel parts of the GeoEye contract, leading to the acquisition of GeoEye by DigitalGlobe in 2013.[6] As a result, DigitalGlobe became the only major U.S. commercial remote sensing provider to the U.S. national security enterprise from 2013 until 2019, when Planet began a service-level contract with the National Reconnaissance Office (NRO).

Current State of Commercial Remote Sensing Market

In recent years, several emerging U.S. companies have been entering the remote sensing market with a plan to leverage improvements in small satellite technology, lower launch costs, and lower manufacturing cost. Compared with Maxar Technologies, which acquired DigitalGlobe in 2017, most of these companies plan to use a larger constellation of smaller satellites. Figure 4.1 shows the estimated number of operating commercial remote sensing satellites from 1995 to 2023, including both U.S. and international companies. Forecasted growth from 2020 to 2023 considers the planned launches by several major companies. But these estimates do not include all providers and therefore may be underestimating the near-term growth (alternatively, missed launch projections by those companies could result in this estimate being an overestimate).

In early 2020, about 225 commercial remote sensing satellites were on-orbit and operating— a dramatic increase over the past decade from the 25 satellites that were operating in 2010. The remarkable and steady increase in the number of operating remote sensing satellites began in 2013. This explosive growth is partly because of the use of larger constellations of smaller satellites by emerging companies relative to longer, established providers (such as DigitalGlobe), which have used smaller numbers of highly capable satellites. About half the growth since 2014 has been made up of Planet Labs' (aka Planet) Dove constellation, which currently includes about 130 nanosatellites.[7] In the near term, companies, such as Capella and BlackSky, are

---

[5] Office of Space Commerce, "ClearView Arrangements Awarded to Three Remote Sensing Firms," press release, March 29, 2003.

[6] Peter B. de Selding, "NGA Letters Cast Cloud over GeoEye's EnhancedView Funding," *SpaceNews*, June 23, 2012; Warren Ferster, "DigitalGlobe Closes GeoEye Acquisition," *SpaceNews*, January 31, 2013.

[7] There is no official consensus definition of satellite mass classes. For the purposes of this report, we use the FAA's definitions, where the prefix *nano* denotes 1.1 to 10 kg, *micro* is 11 to 200 kg, *ini* is 201 to 600 kg, *small* is 601 to 1,200 kg, *medium* is 1,201 to 2,500 kg, *intermediate* is 2,501 to 4,200 kg, and *large* is 4,201 to 5,400 kg. Other mass classes (both smaller and larger) are typically not relevant to commercial remote sensing (FAA Commercial Space

Figure 4.1. Estimated Number of Commercial Remote Sensing Satellites, 1995–2023

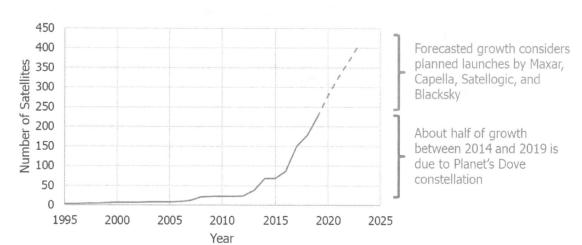

Forecasted growth considers planned launches by Maxar, Capella, Satellogic, and Blacksky

About half of growth between 2014 and 2019 is due to Planet's Dove constellation

SOURCE: RAND analysis of open-source reporting.

planning to expand constellations from several currently operating pathfinder satellites to operational constellations numbering in the dozens of satellites.

In the past decade, global revenues for commercial space-based remote sensing companies have approximately doubled, from $1 billion to $2.1 billion between 2009 and 2018,[8] as shown in Figure 4.2 (solid blue line and left-hand vertical axis), indicating a growth in overall demand for satellite remote sensing. However, because of the dramatic increase in the number of satellites, the revenue per satellite (dashed red line and right-hand vertical axis) has fallen from $50 million to $10 million in the same time frame.

The fall in revenue per satellite could be an indication that new supply has not been fully met by increased demand and a signal of financial risk to the market and individual systems, but there are several other factors to consider and monitor in the future. First, with the increased dependence on smaller satellites and lower launch and manufacturing costs, the required revenue per satellite for commercial success is decreasing. For example, a large and highly capable WorldView satellite would require more revenue to sustain than a nanosatellite from Planet. The second factor is that, although increased supply of commercial space-based imaging may be leading demand, it does not mean that demand will not catch up and support emerging commercial remote sensing companies in the future. For these reasons, the decline in revenue per satellite shown in Figure 4.2 does not demonstrate that companies or systems will necessarily fail commercially. Nevertheless, it illustrates that the commercial remote sensing market is changing rapidly. Uncertainty in the future size of the commercial remote sensing market (with

Transportation and the Commercial Space Transportation Advisory Committee, *2015 Commercial Space Transportation Forecasts*, Washington, D.C.: Federal Aviation Administration, Office of Commercial Space Transportation, April 2015, p. 56).

[8] Summaries from Satellite Industry Association reports from 2010 to 2019.

implications for diversity and number of suppliers) and the financial viability of individual companies or systems introduce commercial risks that must be considered for any government acquisition of space-based sensing.

Figure 4.2. Commercial Satellite Remote Sensing Global Revenue, 2009–2018

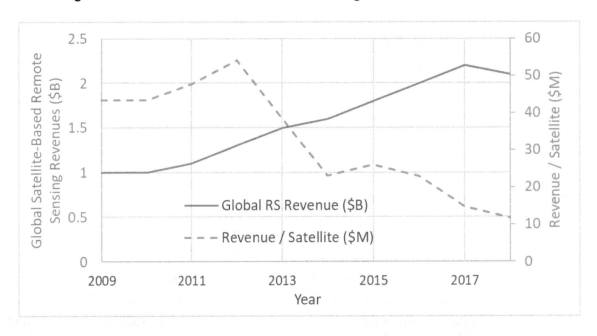

SOURCE: State of the Satellite Industry report summaries, 2010 to 2019 (global satellite-based remote sensing revenue); RAND analysis of open-source reporting (number of operating remote sensing satellites).
NOTE: Falling revenues per satellite does not indicate emerging companies will fail financially, primarily because many emerging operators use smaller and less expensive satellites. See more detailed discussion in this chapter.

U.S. government engagement with the commercial remote sensing market changed in 2017. In 2017 and 2018, the NRO took responsibility for acquiring commercial ISR for the intelligence and defense communities. NRO plans to leverage commercial imagery when possible, in accordance with National Security Presidential Directive 27 and out of a desire to increase capacity.[9] In 2019, Troy Meink, then the director of the Geospatial Intelligence Systems Acquisition of the NRO, said that commercial imagery "is going to give us more capability for the dollar than we would have otherwise."[10] Table 4.1 shows the commercial remote sensing contracts issued by NRO thus far. It includes operational acquisition contracts to Maxar and, more recently, Planet, as well as a range of study contracts.

---

[9] White House, 2003.

[10] Nathan Strout, "How the NRO Learned to Stop Worrying and Love the Commercial Imagery," *Air Force Times*, June 4, 2019a.

Table 4.1. Commercial Geospatial Intelligence Contracts by NRO, as of March 2020

Supplier	Contract Date	Contract Type	Product Type
Maxar	11/6/2018	3-year operational	WorldView-1 to WorldView-3 pan/ satellite/infrared access, image library
Maxar	6/3/2019	6-month study	"Capabilities beyond EnhancedView follow-on"
Planet	6/3/2019	6-month study	Persistent panchromatic/satellite Earth observation
BlackSky	6/3/2019	6-month study	Color Earth observation
HySpecIQ	9/23/2019	Study	Hyperspectral images
Planet	10/15/2019	Multiyear operational	Persistent panchromatic/satellite Earth observation
Capella	12/11/2019	Study	Synthetic aperture radar imaging
Hawkeye 360	12/11/2019	Study	Radio frequency ID and geolocation

SOURCE: RAND analysis of open-source reporting.

Trends in Remote Sensing Technology and Capabilities

New entrants to the commercial space-based remote sensing market have been driving a change in capabilities. Figure 4.3 shows the revisit rate (i.e., temporal resolution), ground-sampling distance (i.e., spatial resolution), and phenomenology for current commercial remote sensing constellations. The figure shows information for the full planned constellation, which may not be complete (e.g., Capella's and BlackSky's).

New constellations, with initial launch in 2014 or later, are mostly clustered toward the top of the plot (higher revisit rate). Proliferated LEO (PLEO) ISR constellations are the most common architecture for new entrants to commercial remote sensing. The spatial resolution of the new constellations is more varied but can be comparable with all but the most capable (and largest satellite) existing systems (e.g., WorldView-4 by Maxar). The increased revisit rate is driven by the larger constellation size of smaller satellites. The spatial resolution can be coarser because of less-capable individual satellites but improving technology and lower manufacturing costs can keep it competitive with previous generation systems.

Synthetic aperture radar (SAR) and hyperspectral phenomenologies, which were rare in commercial providers, are entering the marketplace with improved capacity, revisit rate, and spatial resolution. Because the information is not publicly available for some companies, the potential satellite capability for tasking, cueing, pointing, and dwelling is not reported here for the various satellite programs. Generally, larger and more-capable satellites by commercial providers have such capability, but small nanosatellite-type platforms may not. Large existing satellites, such as later-generation WorldView satellites, are likely to be more capable in these capacities than PLEO constellations of small satellites or nanosatellites, such as the Planet Dove constellation.

Figure 4.3. Capabilities of Commercial Remote Sensing Constellations

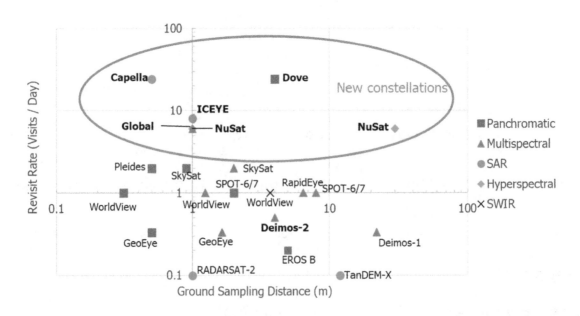

SOURCE: RAND analysis of open-source reporting.
NOTE: This figure includes only constellations with some on-orbit satellites, but information is for the full planned constellation, which may not yet be complete (e.g., Capella's and BlackSky's). Constellations are plotted twice when they use multiple phenomenologies (e.g., panchromatic and multispectral). Constellations with initial launch in or after 2014 are bolded.

Indicators to Watch

The commercial remote sensing market is changing rapidly, and, because of market uncertainty, developing a detailed long-term acquisition strategy for the U.S. national security community may not be prudent at this time. This represents a challenge in the form of requiring long-term monitoring of the market. But it also provides an opportunity to rapidly shift service acquisitions and help shape the emerging segment of the market. Both programmatic and financial indicators may help the U.S. national security community evaluate the state of the market over time.

Some programmatic indicators of a commercial remote sensing provider's financial soundness, operational effectiveness, ability to meet expectations of capability, and ability to build a customer base include

- meeting launch schedule

- achieving successful satellite operations, including orbiting, initial communications, and continuing operations of the satellite
- making services commercially available.

Some indicators that show decreased risk that the company will leave the market because of financial reasons and can be a long-term provider to the U.S. national security community include

- establishing a solid customer base and commercial demand
- maintaining capital.

Key Company Assessments

Established commercial remote sensing companies include Maxar and Planet, because they have fully operational constellations and an established customer base, including service-level agreements with the U.S. government. Emerging companies have partial or no operational systems and need to establish a customer base more fully.

Maxar Technologies is a well-established company with diverse revenues and a long history (through its subsidiary DigitalGlobe and predecessors) of commercial remote sensing for the U.S. national security enterprise. Their WorldView system of three satellites (WorldView-4 failed in 2019[11]) is the current workhorse supplying commercial imagery to the U.S. defense and intelligence communities. The WorldView Legion constellation is the successor system planned for launch in 2021. By using six planned satellites, a reduced revisit time is planned.

Planet, founded in 2010, operates two fully operational satellite constellations and serves an array of customers.[12] Planet spearheaded the PLEO remote sensing constellation through its Dove (aka Planetscope) constellation of nanosatellites. They were initially funded largely through venture capital. Planet initially focused on civil and commercial customers but has been growing its relationship with the U.S. national security community through study and procurement contracts with the NRO.

Other companies are relative newcomers to the commercial remote sensing market. Capella Space is in the process of launching a constellation of 36 SAR microsatellites. If successful, this will represent the first constellation of SAR satellites by a U.S. company, and it will have a significantly larger constellation size than other providers'. BlackSky, a business unit at Spaceflight Industries, is building a constellation of electro-optical/infrared (EO/IR) sensing satellites and highly focuses on building analytic capabilities (e.g., artificial intelligence[13]) and

---

[11] *SpaceNews* staff, "DigitalGlobe Loses WorldView-4 to Gyro Failure," *SpaceNews*, January 7, 2019.

[12] Planet announced that their third constellation, RapidEye, would be retired in March 2020 (see Martin Van Ryswyk, "RapidEye Constellation to Be Retired in 2020," Planet, January 16, 2020).

[13] "BlackSky's global persistent monitoring services combine the state-of-the-art in artificial intelligence, cloud computing, multi-sensor data fusion, activity analysis, and autonomous satellite tasking to rapidly deliver essential alerts to those who need to know" (BlackSky, "Products and Services," webpage, undated).

partnerships with other remote sensing providers, such as Hawkeye 360.[14] Hawkeye 360 is building the first commercial space-based sensing constellation for detecting and mapping radio frequency (RF) spectrum emissions. HySpecIQ is also investigating the application of new phenomenologies to the commercial sector by studying the use of hyperspectral imaging. These emerging companies are venture capital funded, growing in their capabilities, establishing a relationship with the U.S. national security community via NRO study contracts, and looking to establish commercial and civil customer bases. Although there is significant uncertainty about the long-term stability of each individual provider, successful transition of these emerging remote sensing providers would bring capabilities very relevant to DoD.

Table 4.2 summarizes additional characteristics of these U.S. remote sensing companies.

---

[14] Annamarie Nyirady, "HawkEye 360 Provides RF Data to BlackSky," *Via Satellite*, June 3, 2019.

Table 4.2. Summary of Select U.S.-Based Remote Sensing Commercial Service Providers

	Maxar		Planet		Capella	BlackSky	Hawkeye 360	HySpecIQ
Company found	1992 as DigitalGlobe		2010 as Cosmogia		2016	2010	2015	2015
Program name	WorldView	WorldView Legion	Dove (aka Planetscope)	Skysat	Capella	Global	Hawkeye 360	HySpecIQ
Constellation description	3 large satellites	6 large satellites	130+ nanosatellites	15 minisatellites	36 microsatellites	56 microsatellites	18 microsatellites	TBD
Program status	Fully operational	Not operational—planned launches in 1Q21	Fully operational—may be further augmented and refreshed	Fully operational	Partially operational—2 demonstration satellites operating	Partially operational—4 satellites operating	Partially operational—3 satellites operating	Not operational
Phenomenology	Panchromatic, multispectral, and SWIR EO/IR	Panchromatic and multispectral EO	Panchromatic and multispectral EO	Panchromatic and multispectral EO	X-Band SAR	Visible, panchromatic, and color EO/IR	RF detection and geolocation	Hyperspectral imaging
Ground sample distance (m)	Pan: 0.31, satellite: 1.24, SWIR: 3.7	0.3	3 to 5	Pan: 0.72, satellite: 1	0.5	1	Not available via open-source reporting	TBD
Minimum revisit time	1 day	36 minutes	1 day	8 hours	2.5 minutes	4 hours	< 1 hour	TBD
Company investors, customers, foreign relations, and other associations	Publicly held U.S. company, significant U.S. government funding, established and diverse customer base, including U.S. government, allied defense, and intelligence, civil, and commercial clients		Privately held U.S. company, venture capital funded, U.S. government funding via NRO study and operational contracts, diverse customer base, including international clients		Private U.S. company, venture capital funded, limited U.S. government funding via NRO study contract, currently limited customer base	Private U.S. company, venture capital funded, limited U.S. government funding via NRO study contract, currently limited customer base	Private U.S. company, venture capital funded, limited U.S. government funding via NRO study contract, partnerships with BlackSky and Airbus, currently limited customer base	Private U.S. company, venture capital funded, limited U.S. government funding via NRO study contract, currently limited customer base

SOURCE: RAND analysis of open-source reporting.
NOTE: Includes only U.S.-based companies. Ground sample distance and revisit time are based on company claims. Revisit time is given for full constellation and may not yet be achieved by partially operational constellations (e.g., Capella's and BlackSky's).

5. Environmental Monitoring

Mission Scope

For the purposes of this analysis, *environmental monitoring* encompasses both terrestrial and space weather missions. The terrestrial environmental monitoring mission is to provide meteorological and oceanographic information—including forecasts—across maritime, air, and land domains that may impact military operations.[1] To facilitate this, large quantities of environmental data are collected using space-based EO/IR sensors and microwave sensors. EO/IR sensors are typically part of imaging systems that sense both visible and infrared light, where the targeted frequency spectrum depends on the phenomenon of interest. The EO/IR sensors used in environmental monitoring typically perform cloud characterization and provide theater weather imagery. Microwave sensors provide information about ocean surface winds, snow depth, soil moisture, tropical cyclone intensity, and sea ice.

The space environmental monitoring mission provides space environmental data to support forecasts, alerts, and warnings to protect space assets, space operations, and the terrestrial operations that depend on environmental knowledge.[2] Space weather sensors monitor phenomena, such as the solar wind, coronal mass ejections, auroras, and ionospheric disturbances. The data that space weather sensors collect include ionospheric density and scintillation, energetic-charged particle characterization, and the electric field. EO/IR, microwave, and space weather sensors have all been considered for this commercial capability assessment.

In addition to the aforementioned sensors, the scope of this review includes value-added services and Global Navigation Satellite System Radio Occultation (GNSS-RO, aka GPS-RO). *Value-added services* are companies that provide analytics on environmental data and produce weather-related products. GNSS-RO has garnered interest because of its benefit in providing more atmospheric profile information to terrestrial weather forecast models. A GNSS-RO satellite will measure the refraction of the signal from a GPS satellite as it passes through the atmosphere, which gives information about temperature and water vapor. It can similarly be used to measure ionospheric electron density, which can inform space weather forecasts.

---

[1] Joint Publication 3-14, 2018.

[2] Joint Publication 3-14, 2018.

Market Overview

Overall, the commercial environmental monitoring market is still developing, with some variation depending on the segment. The main barrier to maturity in the commercial environmental monitoring market is the high capital expenditure requirement to launch a fully operational satellite constellation. Additionally, the uncertainty in market demand from governments and nongovernmental customers leaves open the question of commercial viability.

The biggest developments have been in GNSS-RO capabilities by PlanetiQ, GeoOptics, and Spire Global, the latter two having launched satellites that produce radio occultation soundings. All are catering mostly to NOAA and DoD as anchor tenants.

The value-added services segment consists of a few startups that have had recent success in raising capital, namely Orbital Insight and Descartes Labs. Both startups are targeting a wide range of customers, focusing in such areas as real estate, agriculture, and energy.

Orbital Micro Systems made headway in the microwave-sensor segment of the commercial environmental monitoring market with the launch of its first satellite in 2019. It still has plans to launch at least six additional satellites.

No companies are currently offering traditional EO/IR sensor services. Similarly, no companies are offering solely space weather-sensor services, except for the ionospheric density data augmented by the GNSS-RO providers.

Trends in the Market—Technology and Capabilities

The value of the commercial environmental monitoring market has been relatively stagnant and potentially shrinking. An estimate of the commercial data market for all Earth observations in 2018 was $1.5 billion, the same value as an estimate in 2012.[3] All Earth observations lump the environmental monitoring and remote sensing markets together.[4] Given the recent growth in the remote sensing sector, it could be that the environmental monitoring sector is becoming smaller.

Within the environmental monitoring market, significant shifts have occurred in the past five years.[5] Orbital Micro Systems launched its first microwave sensor satellite in 2019. The hyperspectral sounding startups, Tempus Global Data and GeoMetWatch, both of which had plans to be hosted payloads on communications satellites in GEO, have lost momentum and have yet to launch. Many launches have taken place in the GNSS-RO segment. Spire Global has continued its early launch success to reach more than 80 satellites on orbit for the CubeSat constellation, producing 5,000 radio occultation (RO) profiles per day. Spire has plans to

---

[3] Euroconsult, *Satellite-Based Earth Observation Market Prospects to 2028*, 12th ed., Paris, 2019; Euroconsult, "First Industry Report on Earth Observation Data Distribution Trends and Strategies," press release, Paris, Montreal, and Washington, D.C., March 27, 2014.

[4] In 2013, the Earth observation commercial data market was made up of 62 percent of very high-resolution optical data, one indicator of remote sensing making up most of the Earth observation market (Euroconsult, 2014).

[5] For greater detail on the state of the commercial environmental monitoring market, see Kim et al., 2016.

continue expanding its constellation. GeoOptics had planned to launch 24 satellites by 2018; however, because of delays, only three of their CICERO satellites are on orbit, producing around 900 RO profiles per day. PlanetiQ had originally planned to have 18 satellites on orbit by 2019; however, its first microsatellite launch experienced a further delay in 2020.

There are notable changes in recent U.S. government demand for commercial environmental monitoring capabilities. One driver for the demand in GNSS-RO data is the uncertainty in the follow-on launch of COSMIC-2, a collaborative GNSS-RO effort with Taiwan. COSMIC-2 successfully launched six microsatellites in 2019, which produce 4,000 RO soundings per day. However, even with the successful launch of COSMIC-2, NOAA has signaled a continued interest in commercial GNSS-RO data.[6]

Both NOAA and DoD have been engaging the commercial environmental monitoring sector through Commercial Weather Data Pilot programs. With its first round of contracts awarded in 2016,[7] the NOAA Commercial Weather Data Pilot has executed two additional rounds of contract awards, with the goal of testing several companies' abilities to provide GNSS-RO data. DoD's Commercial Weather Data Pilot consists of an independent assessment of the utility of commercial GNSS-RO data. For example, a pilot will evaluate the quality of the electron density data taken from the GNSS-RO data by analyzing the effect it would have on existing ionospheric models.[8] Government investment continues to be a crucial source of capital for commercial environmental monitoring companies. The final outcome of the Commercial Weather Data Pilots will be key in determining further government funding for GNSS-RO data.

The U.S. government has recently expressed an interest in engaging the commercial sector for the space weather mission.[9] The GNSS-RO companies offer information about ionospheric density; however, the other aspects of space weather do not currently have a commercial vendor. It will be worth noting whether government interest results in more entrants into the commercial space weather market in the near future.

Regarding new environmental monitoring technology, academic labs, such as MIT Lincoln Laboratory, have focused on improving the sensors—microwave sounders, infrared imaging—and plan to launch them on small satellites.[10] In the future, these could evolve into new startups. Additionally, the value-added services segment has expanded, with business models leveraging AI/ML tools to analyze EO data.

---

[6] Debra Werner, "NOAA Signals Strong Appetite for Radio Occultation," *SpaceNews*, January 15, 2020.

[7] John J. Pereira, Robert Atlas, Joanne Ostroy, William J. Blackwell, Thomas S. Pagano, Jacob Inskeep, and Mark Seymour, "NOAA's CubeSat-Related Activities for Gap Mitigation and Future Planning," briefing, 31st Annual Small Satellite Conference, Logan, Utah, August 8, 2017.

[8] Ralph Stoffler, "United States Air Force Space Weather," briefing, Washington, D.C.: Headquarters, U.S. Air Force, May 2, 2017.

[9] Dan Leone, "NOAA Told to Consider Commercial Data in New U.S. Space Weather Strategy," *SpaceNews*, November 4, 2015.

[10] Pereira et al., 2017.

Key Company Assessments

Many of the key indicators for commercial environmental monitoring options relate to the viability of the company. For instance, a large barrier to entry has been raising enough funding to launch a PLEO constellation. A key metric to watch for in the commercial environmental monitoring sector is a company's constellation status—whether it has launched any satellites and the size of a company's constellation. As noted earlier, the current market is narrow in terms of its focus on GNSS-RO and is not mature in terms of establishing a diverse customer base to ensure profitability. Thus, it is important to track what specific environmental monitoring missions are supported by each company and their current customers or contracts signed. Other metrics to watch include geographic coverage, data accuracy, ground segment infrastructure, and the capital raised by the company.

For the GNSS-RO capability, the major companies are Spire Global, GeoOptics, and PlanetiQ. Some of the key values for each company are shown in Table 5.1. Spire Global has been able to secure significantly more funding and launch many more satellites for its constellation than any other company. It has ten times the output as its nearest competitor, GeoOptics. NOAA has stated that its goal is to have 20,000 GNSS-RO soundings per day for terrestrial weather forecasting.[11] Although Spire Global currently provides more than 5,000 GNSS-RO soundings per day, it was planning to increase this number to 20,000 GNSS-RO soundings per day by the end of the 2020. PlanetiQ, despite launch delays, plans to have 20 satellites on orbit by 2022, with hopes to quickly scale up to 50,000 GNSS-RO soundings per day.[12] If all companies are able to reach their output goals, the supply will likely outweigh the government demand, heightening the competition even more, unless greater commercial demand is realized.

The startups that are pursuing other types of environmental monitoring sensors are relatively isolated. Orbital Micro Systems is the most notable of these with its first CubeSat launch in 2019, which collects microwave radiometer data. It has been boosted by securing $750,000 in Small Business Innovative Research (SBIR) funding in 2019.

The value-added services business segment of environmental monitoring is starting to develop, taking advantage of advances in data science to create software to analyze environmental data. Currently, the big players in the market are Orbital Insight, Descartes Labs, and SpaceKnow, which have raised $78.7 million, $38.3 million, and $5.5 million in capital, respectively. This segment may have the most appeal to nongovernmental customers, because the value-added service providers have targeted such industries as real estate, energy, and agriculture.

---

[11] Jeff Foust, "Acting NOAA Leader Stresses Importance of Public-Private Partnerships," *SpaceNews*, January 22, 2020a.

[12] Werner, 2020.

Table 5.1. Status of Major GNSS-RO Companies

	Spire Global	GeoOptics	PlanetiQ
Year founded	2012	2005	2012
Constellation status	Active (80 CubeSats on orbit), more planned	Active (3 CICERO satellites on orbit)	Planned first launch into polar orbit in 2020 delayed, plans 18–20 on orbit by 2022
Current output	5,000 profiles/day	500 profiles/day	None
Capital raised	$120 million	$5.2 million	$23.9 million
Ground segment	Partner with Amazon Web Services (AWS) Ground Station	Partner with Tyvak Nano Satellite Systems	Kongsberg Satellite Services (KSAT) and Atlas ground stations

SOURCES: Crunchbase, "GeoOptics," webpage, undated; GeoOptics, "GeoOptics Celebrates Two Years of On-Orbit Operations," press release, January 15, 2020; Caleb Henry, "Fresh $18.7 Million Funding Round Puts PlanetiQ Weather Constellation Back on Track," *SpaceNews*, July 11, 2019c; Spire Global, "Spire Taps AWS Ground Station to Extend Ground Station Network," press release, Las Vegas and San Francisco, November 27, 2018; Werner, "Lofty Aspiration for Spire's Weather-Watching Cubesats," *SpaceNews*, September 17, 2015; Werner, 2020.

38

6. Space Domain Awareness

Mission Scope

SDA encompasses the detection, tracking, and characterization as a threat or nonthreat of objects in space. It provides the requisite foundational, current, and predictive knowledge and characterization of space objects and the operational environment on which space operations depend.[1] It is dependent on the integration of space collection, surveillance, data processing, and dissemination. By helping to ensure the survival of space assets and the ability to exercise command and control over them, SDA is an enabler of all other space missions.

Historically, DoD has been the world's provider of the most accurate and comprehensive information on SDA, previously referred to as *space situational awareness*.[2] The U.S. military's Space Surveillance Network, operated by U.S. Space Command through its Combined Space Operations Center, is a global network of sensors that are used to detect, track, and monitor artificial objects in Earth's orbit. Data from these sensors are processed and used by DoD operators and provided for free to other satellite operators for the purposes of space traffic management (STM).[3]

The space operating environment is changing rapidly. STM is challenged by an increasing number of objects and operators, particularly in LEO. Both active artificial satellites and orbital debris (e.g., discarded LV stages, dead payloads, debris from past collisions) can pose threats to satellite operations. As the number of satellites and amount of space debris has increased over time, it has become more challenging for satellite operators to have accurate, comprehensive, and timely data for STM functions.

The Space Surveillance Network is based on hardware that was developed to provide a missile-warning capability and was not optimized to detect, track, and monitor thousands of objects that are in Earth's orbit. The recent reestablishment of the USSF, which was declared operational in March 2020,[4] will drastically increase the capacity of the Space Surveillance

---

[1] See Joint Publication 3-14, 2018. Various definitions of SSA and SDA exist. For a collection of some published definitions, see Bhavya Lal, Asha Balakrishnan, Becaja M. Caldwell, Reina S. Buenconsejo, and Sara A. Carioscia, *Global Trends in Space Situational Awareness (SSA) and Space Traffic Management (STM)*, Washington, D.C.: IDA Science and Technology Policy Institute, IDA Document D-9074, April 2018, Appendix B.

[2] SDA replaced SSA as the preferred term within the Air Force per an October 4, 2019, memo by Maj Gen John Shaw, then–Deputy Commander of Air Force Space Command (see Sandra Erwin, "Air Force: SSA Is No More; It's 'Space Domain Awareness,'" *SpaceNews*, November 14, 2019b).

[3] *Space traffic management* can be defined as safe access into outer space, operations in outer space, and return from outer space to Earth. Most commonly, it involves the prediction and avoidance of potential collisions (aka conjunctions) between space objects but can also include other aspects, such as management of electromagnetic spectrum emissions. Additional published definitions are listed in Lal et al., 2018, Appendix B.

[4] Sandra Erwin, "Space Fence Surveillance Radar Site Declared Operational," *SpaceNews*, March 28, 2020c.

Network, particularly in LEO. However, "operators increasingly view today's DoD SSA system and service as inadequate to achieve safe operations in space,"[5] particularly for sensitive operations, such as formation flying. Commercial suppliers of SDA observations, tracking, and analysis may be able to better meet the needs of satellite operators in some cases.

Space Policy Directive-3, signed in June 2018, shifted the responsibility of providing an "open architecture SSA data repository"[6] to the U.S. Department of Commerce, as noted in the publicly releasable portion of the DoD catalog. In addition to other principles intended to improve safe access to an increasingly congested space domain,[7] this directive will result in the transition of public- and commercial-facing STM functions away from DoD. In addition to DoD sensors, which are expected to continue to be operated by DoD, the SSA data repository may be supplemented by commercial or other sensors.[8] However, challenges, particularly lower-than-requested funding appropriations,[9] have made the time frame for this transition unclear, and this responsibility continues to primarily fall on DoD.[10]

In addition to supporting STM for safe access to space for military, civil, and commercial uses, SDA also encompasses many military-specific tasks and missions. These missions are becoming increasingly important, because space is seen not as a benign environment housing supporting capabilities, but because it is a contested "warfighting domain just like air, land and sea."[11] SDA supports the National Space Defense Center's[12] ability to "rapidly detect, warn, characterize, attribute and defend against threats to our nation's vital space systems."[13] Space battle management will require more than relatively static tasking and catalog maintenance, including frequent or dynamic observations of certain objects, timely and actionable indications and warning of potential threats, maintenance of custody of known potential threats, access to intelligence, detailed characterization of satellite capabilities, and integration between different

---

[5] Lal et al., 2018, p. iii.

[6] White House, "Space Policy Directive-3, National Space Traffic Management Policy," presidential memorandum, June 18, 2018.

[7] For example, by creating minimum safety standards and global promotion of best practices and protocols.

[8] For example, DoD has shared its experimental Unified Data Library (UDL), which collects observations from various DoD and commercial sources of SDA data, with the U.S. Department of Commerce (see Theresa Hitchens, "Crider: SSA Data 'Library' Will Open to Allies," *Breaking Defense*, May 3, 2019a).

[9] Theresa Hitchens, "Hill Nixes Trump Space Tracking Plan," *Breaking Defense*, December 19, 2019b.

[10] Rachel S. Cohen, "Space Traffic Transition to Commerce Hits Speed Bumps," *Air Force Magazine*, February 14, 2020.

[11] Gen John "Jay" Raymond, as Commander of U.S. Air Force Space Command, now Commander, U.S. Space Command and Chief of Space Operations, U.S. Space Force (see Colin Clark, "Exclusive: War in Space 'Not a Fight Anybody Wins'—Gen. Raymond," *Breaking Defense*, April 6, 2017).

[12] One of two subordinate commands to U.S. Space Command.

[13] Shellie-Anne Espinosa, "National Space Defense Center Transitions to 24/7 Operations," Air Force Space Command (Archived) webpage, January 26, 2018.

platforms. Some SDA sensors should be treated consistently with the role of space ISR weapon systems, a role that they must now fulfill.

Market Overview

Several commercial companies are emerging that have the goal of providing SDA observational data and analysis products to the national security enterprise and commercial customers. These companies operate ground-based sensing technologies, including optical telescopes, radar, and passive RF sensing and are currently operating sites around the world. This industry, particularly the segment focused on acquiring observational data instead of software, remains in its early stages, with clear growth but potentially significant uncertainty relative to the future size and stability of the industry.

Like other space service mission areas, commercial SDA providers hope to provide services to both government and commercial customers. Malfunctions, whether because of collisions, poor or mischaracterized orbital placement, or operational interference (e.g., RF spectrum interference) can have large effects on the revenues of commercial space operators, particularly those that use large and expensive satellites. One recent analysis has estimated a cumulative potential revenue loss as a result of satellite malfunction by commercial space operators to be about $16 billion between 2019 and 2028.[14] This creates a potential market for commercial SDA providers to build a customer base of commercial space operators.

Because the U.S. government provides some STM data publicly and for free, the commercial business case for SDA can be challenging, and commercial SDA providers must add value beyond the government-provided data. For higher-risk operations (e.g., proximity operations and maneuvers), commercial SDA providers may be able to offer tailored and, in some cases, more-accurate information to space operators and add value beyond government-published data. However, the nascent nature of the commercial SDA industry and hesitation on the part of space operators to pay for STM information when the government provides data for free create challenges for closing the commercial business case for SDA companies.

Every commercial SDA provider whom we interviewed expected that the U.S. national security enterprise would be the anchor customer in the near-term and potentially indefinitely. Several of those providers stated that they would be unable to sustain their current SDA observational capabilities without business or other funding support from the U.S. government. This creates risk for the government by requiring continuous support for some commercial SDA providers if the government would like to guarantee continued commercial SDA service availability. Funding uncertainty can cause stagnation in the deployment of additional commercial SDA capabilities or even drive companies or services out of the market. However, the dependence on government support positions the government to potentially shape the market

---

[14] Dallas Kasaboski, "Satellite Social Distancing," Northern Sky Research website, March 31, 2020.

and individual SDA providers to fulfill the needs of U.S. national security space operators. Even in this situation, commercial customers can help to stabilize the SDA industrial base and share the capital costs of SDA infrastructure with the U.S. government.

DoD has been taking concrete steps to begin business relationships with commercial SDA providers. Space and Missile Systems Center (SMC)[15] and the Air Force Research Laboratory have spearheaded these efforts with the integrated Commercial Augmented Mission Operations (CAMO) program and the development of the UDL. These experimental programs have provided funding to several SDA companies to provide observational data and develop systems and processes that fuse data from various commercial and government sources in a single repository and distribute that repository to government space operators.

Trends in the Market—Technology and Capabilities

Commercial SDA services can include observational systems that collect data and software-based capabilities that conduct analysis using those observations. Commercial SDA data providers have deployed optical, radar, and passive RF-sensing capabilities and established observational locations globally. ExoAnalytic Solutions, Numerica, LeoLabs, and Rincon have previously provided commercially acquired data to DoD space operators via experimental programs, such as CAMO.

Commercial SDA technology is technically mature. Optical space telescopes, radar for space sensing, and RF-based location of electromagnetic emitters are established technologies and have long contributed to the U.S. space object catalog. Technological risk is not expected to be a primary driver of risk for commercial SDA. Although commercial providers may not be introducing novel sensing capabilities, they have other advantages. For example, commercial SDA operators may be able to establish sensing sites in locations that DoD would not because of political or security concerns.

In addition to owning and operating sensors, commercial SDA providers have developed software capabilities to identify satellites, predict the trajectory of observed satellites, conduct assessments for STM and other missions, and distribute information. In most cases, software tools are designed to be complementary with the company's data collection.[16] In some cases, commercial SDA companies operate their own space operations centers. Several have been operating for years, including the Space Data Center and the Commercial Space Operations Center, both operated by Analytical Graphics, Inc. Providers of observational data have been introducing their own operations centers more recently. For example, ExoAnalytic Solutions has stood up their ExoAnalytic Space Operations Center.

Although the availability of commercial software that schedules observations, tasks sensors, collects data, analyzes, and distributes processed information is not new, the number of potential

---

[15] SMC has changed its name to Space Systems Command as of late 2021.

[16] There are some exceptions, such as Analytical Graphics, Inc., that focus solely on software-based capabilities.

options has increased with the number of commercial providers. This can be advantageous, giving more options and competition within the market that could potentially be leveraged by DoD. However, the increasing number of data providers and analysis software increases the challenge of achieving interoperability among commercial providers and between commercial providers and DoD. The CAMO and UDL programs have been developing one method of overcoming the interoperability challenges for several SDA providers.

Key Company Assessments

In this section, we describe the characteristics of several emerging U.S. companies that collect observations of space objects (see Table 6.1). They focus on data-collection capabilities, but all these companies also have some level of software-based capability for data processing and distribution. The most common type of sensor used by commercial SDA providers is optical telescopes. ExoAnalytic Solutions, Numerica, and L3Harris all depend on observations from this type of sensor.

With more than 300 telescopes distributed throughout more than 25 locations, ExoAnalytic Solutions has the largest commercial system of ground-based SDA optical telescopes in the world. ExoAnalytic began with a focus on data collection but has also made some software tools available and operates a space operations center. Although it has commercial customers, particularly for maneuvers and proximity operations in GEO, government demands drive its large number of telescopes; this would likely not be sustainable with only commercial customers. Several business models are possible, from recurring data subscriptions to near-real-time control and data ingestion through telescope-leasing models.

Numerica began with a focus on software development, particularly their MFAST tool, which leverages multiple sensors to rapidly produce orbit characteristics and perform processing for events, such as uncorrelated tracks or closely spaced objects. Building on MFAST, Numerica has developed a suite of software to handle everything from scheduling and tasking of automated telescopes to the distribution and visualization of data products. More recently, Numerica has been building a global network of optical telescopes, including a demonstration of making observations during the daytime.

Table 6.1. Commercial SDA Ground-Based Observational Data Providers

Company Name	Phenomenology	Sites/Sensors	Target Domain(s)
ExoAnalytic Solutions	Optical	30/300+	GEO, MEO, HEO
Numerica	Optical	18/130+	GEO, MEO, HEO
L3Harris	Optical	Not available	GEO
LeoLabs	Radar	3/3	LEO
Kratos	Passive RF	21/>80	GEO
Rincon	Passive RF	11 / 21	GEO, MEO, HEO, limited LEO

SOURCE: RAND analysis of open-source data.

LeoLabs is the only commercial SDA provider that uses radar sensors. It currently has three phased-array radars operating, with plans for expansion. Two of those radars are in ultra high frequency (UHF) and one in S-band (allowing detection of objects down to 2 cm[17]). The use of phased-array radars, compared with more-prolific optical telescopes, gives a distinct advantage in achieving comprehensive and persistent coverage of the lower orbital altitudes, including LEO.

The final sensor type currently used in commercial SDA is the passive measurement of RF emissions from satellites. Kratos is a large aerospace contractor with a space division that primarily focuses on satellite command and control and signals-interference monitoring for both commercial and DoD customers. Kratos has been in the business of signals monitoring for about 15 years and is planning to offer passive RF SDA to government and commercial organizations internationally. Rincon is a company that has traditionally focused on signal-processing hardware and software for defense customers and has more recently introduced commercial SDA services. Compared with other sensor types, RF detection can rapidly and accurately characterize satellite maneuvers but does not have the ability to detect non–RF-emitting objects, such as space debris or threat satellites that have "gone dark." Rincon does not plan to sell their SDA data to purely commercial customers but would be willing to engage in various business models with the national security enterprise, including a service agreement similar to other commercial providers or a model closer to more typical DoD acquisitions.

---

[17] LeoLabs, "Global Phased-Array Radar Network," webpage, undated.

7. Data Transmit/Receive Networks

Mission Scope

In this chapter, we focus specifically on the ground antenna network that performs transmit and receive functions with spacecraft. Such a data-transport network can be used for mission data dissemination (e.g., downlink remote sensing data), launch support, and satellite telemetry, tracking, and control. Ground stations that are available for use on the commercial market are called *ground station as a service* (GSaaS). The U.S. military currently conducts transmit and receive operations primarily using the AFSCN.[1] Our focus on this specific mission was guided by particular interest expressed by the project sponsor. The review of commercial capabilities in this area includes ground antennas with functionality in any spectral band. We limited this review to companies that are U.S.-based or have a U.S. subsidiary, because these are limiting factors for doing business with DoD. However, one company, KSAT, is included despite not meeting the criteria, because it has established previous relationships with NASA.

Market Overview

The GSaaS market is growing and maturing, spurred by the recent growth in proliferated LEO remote sensing satellite constellation launches. Six companies have built at least an initial segment of their ground station network: Atlas Space Operations, Amazon Web Services (AWS) Ground, KSAT, RBC Signals, Swedish Space Corporation (SSC), and Viasat. All provide services to satellites in LEO, and some also offer MEO and GEO services. Evident by the customer base, the demand is increasing primarily because of the launch of commercial Earth-observation satellite constellations[2]; companies opt not to build their own custom ground stations. A secondary driver is government demand—for instance, the growing demand for AFSCN antennas to transmit and receive data requires a decision to either build more organic capacity or augment the existing network with commercial capabilities. Some government users are already leveraging commercial ground stations for data transfer, including NASA and DoD.[3]

---

[1] After the completion of this research in September 2020, the name of AFSCN was changed to the Satellite Control Network.

[2] See earlier chapters of this report on remote sensing and environmental monitoring for more discussion on the growth of the commercial space market.

[3] Air Force Research Laboratory, "Commercial Augmentation Service (CAS)," briefing, February 4, 2020, Not available to the general public.

Competition to offer GSaaS to newly proliferated LEO commercial satellite operators has grown. The market has matured quickly in the last five years as Atlas, AWS Ground Station, and RBC Signals have set up operational global ground station networks. Furthermore, KSAT and SSC, which have been around for more than a decade longer, have begun adjusting their networks to accommodate the trend in proliferated LEO.

The emerging technological developments of phased-array antennas and optical communications are some factors that provide diversity among GSaaS companies.[4] These new technologies could improve GSaaS capability in the future. The phased-array antenna will be able to handle more than ten simultaneous connections to satellites in LEO and reduce the number of antennas needed. AWS Ground Station has recently teamed up with Lockheed Martin to consider a shift to phased-array antennas. Atlas Space Operations has already deployed a phased-array antenna in New Mexico and is prototyping another one for the Air Force. SSC has also publicly expressed interest.[5] Although none of the GSaaS companies with operational networks use optical communications, it is worth watching when its once-prohibitive cost starts to decrease and data-transfer capacity demands increase. NASA has completed a demonstration of an optical link to a satellite in LEO. The startup BridgeComm is working with SSC to build a ten-station network of laser-equipped ground stations, and it is targeting remote sensing companies as future customers.[6]

Going forward, it will be important to watch which companies sign contracts with government users and major commercial customers. Many of the companies are hoping to sustain some business from remote sensing operators as a customer base, however, that sector is still evolving, with startups still competing for financial viability, as discussed in Chapter 4. DoD has recognized a growing need for ground station data-transport capacity from the AFSCN and has, in turn, started to engage the commercial sector. As mentioned earlier, the Air Force has sponsored a prototype of a phased-array antenna from Atlas Space Operations. The Air Force has also supported a SBIR contract with RBC Signals to optimize scheduling for the ground network, and the Air Force Research Laboratory Commercial Augmentation Service project has also been engaging commercial ground station providers through SBIR grants.

One issue for the newer entrants into this market area is the regulatory barriers associated with transmitting to a large number of spacecraft. The International Telecommunications Union regulates satellite transmissions such that a company needs a license for every satellite to which it wants to transmit. Furthermore, each country governs the satellite transmissions within its own

---

[4] Also called *electronically steered antennas*.

[5] Caleb Henry, "Lockheed Martin Mulls Electronically Steered Antennas for Verge Ground Station Expansion," *SpaceNews*, May 8, 2019a.

[6] Debra Werner, "Are Laser Links Ready for Prime Time?" *SpaceNews*, May 22, 2019; Caleb Henry, "Commercial Laser Comm Edges Closer to Reality," *SpaceNews*, June 26, 2018.

domain, which must be considered when contemplating a global ground station network. The licensing will become a bigger burden as GSaaS providers take on an internationally diverse client base.[7]

Key Company Assessments

The most important indicators for assessing GSaaS companies are the ground station network status, the frequency bands available, and the types of customers they are pursuing. Measures that describe the ground station network status include the number of operational ground stations and antennas, near-term plans for additional ground stations or antennas, and the geographic locations of the ground stations. In combination, the number and location of the antennas can speak to the overall network capacity and which orbits the network can support. The different supported orbits can influence the types of customers and which specific space missions the company can attract. An important consideration with frequency band availability is that many DoD satellites require a transmit frequency in the L-band. Most commercial providers do not have L-band transmit service because commercial satellite operators do not use the L-band frequency. The types of customers that each company is pursuing can highlight whether these competing companies are going after the same customers or whether there is sufficient diversity to support all competitors in the market.

Some of the key data values of the main GSaaS companies—Atlas Space Operations, AWS Ground Station, RBC Signals, BridgeComm, SSC, Viasat, and KSAT—are shown in Table 7.1. Of the networks that are operational, KSAT and RBC Signals have by far the largest number of ground stations. KSAT has the advantage of being established earlier, although it only turned its attention to the small satellite market in 2016. RBC Signals has a large network because of its business model, which leverage the excess capacity of existing networks.[8] All networks offer services to satellites in LEO because of the demand signal from remote sensing proliferated LEO constellations. Regarding geographic distribution, current ground station locations are shown in Figure 7.1. A notable distinction is that KSAT has the best polar coverage, with the only station in Antarctica and the Arctic. This is a distinction that many other companies are looking to close in on with future planned ground station sites. One other major distinction is the coverage over Russia and most of Asia, which is dominated by RBC Signals.

Looking at current customers and contracts (bottom of Table 7.1), it is unclear whether the U.S. government or commercial customers is the main target. The companies that have been around the longest—SSC, Viasat, and KSAT—notably serve government clients. SSC and

---

[7] Mike Carey, "Why the 'Amazonification' of Satellite Data Communications Is a Good Thing," *SpaceNews*, December 11, 2018.

[8] RBC Signals has its own antennas but has also built partnerships with existing ground station operators with excess capacity on their antennas, which they sell to other customers.

KSAT contract with the United States and foreign government space agencies. Their history could be an advantage for future government business. Between the newer entrants to the market (i.e., Atlas, AWS Ground Station, and RBC Signals), RBC Signals appears to target the U.S. government the most and has benefited from two SBIRs awards, one of which was sponsored by the Air Force. Atlas is actively engaging with DoD, NOAA, and NASA, although it has some diversity with commercial remote sensing customers as well. Current AWS Ground Station customers are all commercial; however, AWS Ground Station does have plans to engage with U.S. government agencies.

The capital raised by Atlas Space Operations, RBC Signals, and BridgeComm is worth tracking because future funding differences could signal differences in commercial viability. AWS Ground Station has some advantage stemming from the well-established and profitable AWS. BridgeComm is leveraging a partnership with SSC, a more established company.

Another factor to keep in mind when considering relationships with the U.S. government is a company's headquarters or subsidiary location and any ties to foreign governments, including funding or supply chain. These may prohibit certain government contracts.

Table 7.1. GSaaS Company Data

	Atlas Space Operations (2017)	AWS Ground (2019)	RBC Signals (2015)	BridgeComm (2015)	SSC (1972); SSC Space U.S. (1996)	Viasat (1986)	KSAT (2002)
Headquarters location	Traverse City, Michigan, USA	Seattle, Washington, USA	Redmond, Washington, USA	Denver, Colorado, USA	Solna, Sweden (headquarters)/ Horsham, Pennsylvania, USA (U.S. subsidiary)	Carlsbad, California, USA	Tromsø, Norway
Service offerings	Global ground station network; data transport and warehousing with AWS Global Cloud Infrastructure	Global ground station network near AWS data storage centers; leverage cloud computing services	Use excess capacity of existing partner networks; dedicated or on-demand stations; distributed computing at station	Planned: global network of optical ground stations	Global, multi-mission ground station network for whole satellite mission life cycle	Secure networking and satellite internet access, small ground station network; sell equipment (e.g., military communications)	Global ground stations; EO services (e.g., oil/ice monitor, vessel detection, ground motion)
Network status	Operational (9 sites; 6 more in development)	Operational (6 sites; more in development)	Operational (50 sites with 70 antennas)	In development	Operational (18 sites)	Operational (two sites, 12 more in development)	Operational (24 sites with 180+ antennas)
Orbits supported	LEO	LEO, MEO	LEO, MEO, GEO	LEO	LEO, polar, GEO	LEO	LEO (polar, inclined, equatorial)
Frequencies	Very high frequency (VHF), UHF, S- and X-bands	S-, X-, and narrowband	VHF, UHF, and S-, C-, X-, Ku-, and Ka-bands, optical	Optical	UHF and S-, X-, C-, Ku-, Ka- and L-bands	UHF and S-, X-, Ku-, Ka-, and L-bands	UHF, VHF, and S-, X-, C-, Ka-, and L-bands
Current customers/ contracts	Prototype for SMC; Space Systems Loral, BlackSky, Helios Wire; NASA; NOAA; DoD (Air Force FalconSat-6)	Spire, Capella, Maxar DigitalGlobe, Myriota, D-Orbit, NSLComm, Open Cosmos, Thales Alenia Space, and more	Two SBIR awards (one with the Air Force), pursuing government customers first but do not currently provide GSaaS	TBD	Government space agencies: German Aerospace Center, European Space Agency, NASA, French Space Agency, Japan Aerospace Exploration Agency	[GSaaS antennas only] DoD, earth observation companies	Government defense and intelligence (Space Norway, NASA, European Space Agency; Rocket Lab; Iceye; SKY Perfect JSAT

SOURCE: RAND analysis of open-source reporting.
NOTE: Shows service offerings and current customers/contracts focusing on GSaaS-relevant information. Some companies offer services related to other missions.

Figure 7.1. Ground-Station Locations for Commercial GSaaS Networks

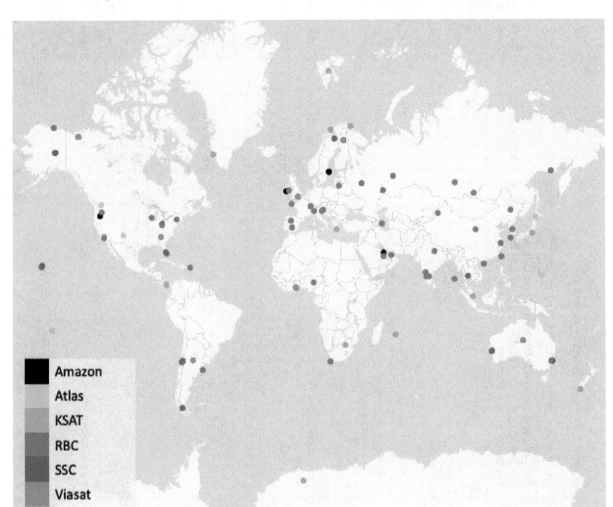

SOURCES: AWS, "AWS Space & Ground Station," briefing, May 27, 2020, Not available to the general public; Atlas Space Operations, "Global Antenna Network," webpage, undated; KSAT, "Ground Network Services," webpage, undated; RBC Signals, "Locations," webpage, undated; SSC, "SSC's Global Ground Station Network," webpage, undated; Viasat, "Real-Time Earth: Rethinking Ground Segment as a Service," brochure, Carlsbad, Calif., 2020.

8. Space Logistics

Mission Scope

The term *space logistics* is loosely defined and includes a range of services that support activities in space. Prominent examples of space logistics are propellent depots, space debris removal, and satellite servicing. The wide range of services in this sector all require similar technologies, namely rendezvous and proximity operations (RPO) and robotics. This chapter focuses on the on-orbit satellite servicing (OoSS) market, one of the most-promising services within the space logistics sector.

Market Overview

The OoSS market is still in the development stage. The full spectrum of capabilities for OoSS includes inspection, capture, repair, replacement, refueling, orbital change, and upgrades of other spacecraft. Various sectors are still developing the technologies required to conduct these capabilities. On the government side, NASA and DARPA are developing servicing vehicles, called OSAM-1 (On-orbit Servicing, Assembly and Manufacturing 1) and RSGS (Robotic Servicing of Geosynchronous Satellites), respectively, capable of refueling and servicing satellites. On the commercial side, companies are developing holistic servicing vehicles, but some are also focusing on developing specific technologies required for the evolution of the sector, such as RPO sensors and robotics. The companies that are developing the market are diversified in size and type, ranging from large government contractors to startups.

In 2019, Northrop Grumman launched the first servicing vehicle under a contract with Intelsat. Its Mission Extension Vehicle 1 (MEV-1) is designed to dock with an Intelsat satellite and extend the life of it by five years by providing orbit maneuvering capability.[1] Although the life extension is not achieved by active refueling or repairing, MEV-1 is the first servicing vehicle launched to open the new commercial market.

Commercial satellite operators have expressed their interest in the possibility of using the OoSS capabilities if and when they become available. Traditionally, the only choice that satellite operators had when an operating satellite was reaching the end of its life was to replace it with a new satellite. With the development of OoSS, satellite operators will be able to weigh the costs

---

[1] Life extension of a satellite can come in various forms, depending on the life-limiting factor of the host satellite. These factors include, but are not limited to, lack of fuel for disposal, lack of fuel for station-keeping, loss of operational capability, and loss of stabilization. A servicing vehicle with a full range of OoSS capabilities, once developed, will be able to perform life-extension services for all life-limiting factors. MEV-1 is designed to provide life extension by allowing the host satellite to be maneuvered without expending its onboard fuel, which can then be used for extended operations.

and benefits involved in servicing versus replacing a satellite, enhancing the ability to manage their fleets as they face fast changes in technology and consumer demand.[2] However, the market has been hesitant to react to this technology for various reasons, including rapid technological advancements in HTS and uncertainties in the traditional SATCOM business.[3]

The satellites in GEO are costly and exquisite.[4] Hence, GSO companies focus on gaining efficiency from the current fleet. Refueling satellites that have outdated technology may not be commercially viable as end users demand more and faster data connection. For the companies that operate in the lower orbits, the life of the satellites is shorter and less exquisite than the satellites in GEO. Therefore, the satellites can be replaced more frequently and be less costly. These are a few examples of the uncertainties in the commercial viability of the OoSS sector that would have to be resolved as the market matures.

Trends in Market—Technologies and Capabilities

In the past few years, the most significant event in the OoSS sector is the successful launch of MEV-1, as just explained. Although the launch of MEV-1 exhibited the potential for commercial use of OoSS technology, the reaction from the market is still mixed. From the national security perspective, the application of OoSS technology is welcomed. A RAND study from 2016 indicated that DoD has potential for shaping this emerging and uncertain market and may play a role to ensure commercial RPO developments do not threaten national security.[5]

In 2019, DoD awarded contracts that created more opportunity for OoSS technology to be applied for national security, along with the ongoing RSGS development by DARPA. SMC awarded a contract to Northrop Grumman's SpaceLogistics to conduct a study on servicing four national security satellites.[6] Furthermore, the Defense Innovation Unit announced an open solicitation for servicing vehicles capable of multi-orbit payload transport and fuel-depot delivery,[7] and the Army is considering the application of OoSS to revive satellites that are not functioning.[8] In this early phase of the OoSS market development with only a few commercial contracts identified, the DoD contracts are signs that the department may become a significant anchor customer for the market.

---

[2] Kim et al., 2016; Sandra Erwin, "In-Orbit Services Poised to Become Big Business," *SpaceNews*, June 10, 2018; Mark Holmes, "Satellite Servicing Becomes an Actual Market," *Via Satellite*, March 2019a.

[3] Sven Eenmaa, "Investment Perspectives: Conferring on On-Orbit Satellite Servicing," *ISS 360: The ISS National Lab Blog*, November 15, 2018.

[4] The term *exquisite* is often used in the satellite industry for satellites that have multiple, highly sophisticated functions and high-performance requirements.

[5] Kim et al., 2016.

[6] Nathan Strout, "The Pentagon Wants to Extend the Life of Satellites and Refuel on Orbit," C4ISRNet, October 1, 2019b. SMC is now called the Space Systems Command as of late 2021.

[7] Nathan Strout, "The Pentagon Wants a Roadside Assistance Service in Space," C4ISRNet, February 5, 2020.

[8] Theresa Hitchens, "SpaceLogistics Sat Servicing Mission Taps New Markets," *Breaking Defense*, March 20, 2020.

From the commercial sector, however, potential clients are maintaining the wait-and-see posture.[9] Commercial satellite operators weigh their investment and replacement strategies based on the benefits, costs, and risks associated with the on-orbit servicing vehicles, along with the current market conditions.[10] For example, the recent decline of GEO satellite orders is seen as a sign for some satellite manufacturers to divest the business and move toward smaller satellites.[11] Unfortunately, for the OoSS, small satellites have shorter lifespans and are more costly to service than to replace.[12] The development of proliferated LEO small satellite constellations also adds to the uncertain outlook for OoSS demand. Another satellite manufacturer had stated that satellite servicing is currently less attractive when considering costs, while communication satellites are getting cheaper to manufacture.[13]

Understanding the mixed reaction, vendors in the OoSS market are considering various types of mission vehicles to provide lower pricing points.[14] Nevertheless, an indicator to pay attention to going forward in the OoSS sector is the number of service agreements and contracts coming from geostationary satellite operators. More contracts between satellite-servicing companies and satellite operators would indicate that the market finds the OoSS technology commercially viable.

Despite the sector being in a very early stage, developments in various technologies, such as RPO, robotics, and spacecraft bus designs, are in progress both in the commercial sector and with the government agencies previously mentioned. These efforts are not just for public or private sectors—public-private partnerships are evident. For example, Maxar is contracted by NASA to demonstrate in-space assembly as a part of the OSAM-1 (formerly known as Restore-L) mission,[15] and Northrop Grumman is developing the spacecraft bus for DARPA's RSGS program.[16] As the companies and the government agencies further develop these technologies, more capabilities, such as refueling, repairing, and upgrading, will be available to provide the full spectrum of satellite servicing, thereby opening more business applications. Furthermore, the technologies being developed in the OoSS sector are likely to be spun off to start new space

---

[9] Mark Holmes, "Satellite Manufacturers Enter a Whole New World," *Via Satellite*, September 19, 2019b.

[10] Kim et al., 2016.

[11] Jeff Foust, "Rethinking Satellite Servicing," *Space Review*, February 4, 2019a.

[12] Joshua P. Davis, John P. Mayberry, and Jay P. Penn, *On-Orbit Servicing: Inspection, Repair, Refuel, Upgrade and Assembly of Satellites in Space*, Arlington, Va.: Center for Space Policy and Strategy, April 2019.

[13] Caleb Henry, "Airbus Impressed by Northrop Grumman, but Remains Undecided on Satellite Servicing," *SpaceNews*, March 11, 2020c.

[14] Holmes, 2019b.

[15] Caleb Henry, "Maxar Wins $142 Million NASA Robotics Mission," *SpaceNews*, January 31, 2020a.

[16] Sandra Erwin, "DARPA Picks Northrop Grumman as Its Commercial Partner for Satellite Servicing Program," *SpaceNews*, March 4, 2020b.

logistics sectors, such as orbital fuel depot, orbital debris removal, and in-space assembly and manufacturing.

Key Company Assessments

As shown in Table 8.1, Northrop Grumman is the only company with an active servicing vehicle currently in orbit. The company also has a follow-on vehicle, MEV-2, that was launched in late 2020 under a second contract with Intelsat.[17] Tethers Unlimited is another company developing OoSS technologies, but for smaller satellites in the lower orbits using microsatellite-class servicing vehicles that can refuel and service. A company based in the United Kingdom, Effective Space, is developing a similar technology using small space drones to provide OoSS capabilities.

Whereas these three companies are designing servicing vehicles that will provide a wide range of OoSS capabilities, other companies are focusing on just a few capabilities. For example, Astroscale, headquartered in Singapore with a subsidiary in the United States, has been developing a small vehicle that assists satellites at the end of their life cycles. The vehicle, which launched in March 2021,[18] has been designed to inspect, dock, and deorbit as an end-of-life service. This type of disposal service allows the satellite operators to maximize the utility of the onboard fuel for extended operations. The company also plans to use the same technology for debris-removal services. OrbitFab is focusing on the refueling aspects of satellite servicing by developing orbital fuel depots. Maxar, a company experienced in a variety of space technologies outside the OoSS sector, is developing technologies, such as robotics, spacecraft bus, and RPO sensors, needed for the evolution to the full range of OoSS activities.

---

[17] Mary Beth Griggs, "Two Commercial Satellites Just Docked in Space for the First Time," *The Verge*, February 26, 2020.

[18] Astroscale, "ELSA-d," webpage, undated.

Table 8.1. Key Space Logistics Companies

	Northrop Grumman	Astroscale	Maxar	OrbitFab	Tethers Unlimited	Effective Space
Active servicing vehicle	MEV-1	Not applicable	Not applicable	Not applicable	Not applicable	Not applicable
Location/ U.S. subsidiary	Virginia, USA	Singapore (headquarters)/ Colorado, USA	Colorado, USA	California, USA	Washington, USA	United Kingdom
OoSS technology focus	End-to-end servicing vehicle	Inspection, re-orbit, and de-orbit	Robotics, spacecraft bus, RPO, sensors	Orbital fuel depot	Small servicing vehicle, robotics, recycling 3D printer	Life extension through orbit control using small space drones
Contracts/ funding in place	Two contracts with Intelsat for MEV-1 and MEV-2	$132 million raised from venture capital; includes a Japanese government–backed fund	Contracts to provide robotics and other parts to NASA's OoSS missions (Restore-L and Dragonfly)	Contracts to supply water to International Space Station; $3 million raised as seed funding	$32.7 million in SBIR grants from NASA and DoD	$15 million raised from venture capital; $100 million deal with a GEO satellite operator
Initial service offering	October 2019	Demo mission in 2020	Not applicable	Not applicable	2022–2023	2020

SOURCES: Northrop Grumman, "SpaceLogistics," webpage, undated; Astroscale, undated; Maxar, "OSAM-1 and Spider," webpage, undated; OrbitFab, "Products & Hardware," webpage, undated; Tethers Unlimited, "In-Space Services," webpage, undated; Arie Halsband, "Pioneering Last-Mile Logistics in Space," Effective Space Solutions, February 2017; SBIR STTR, "Tethers Unlimited, Inc.," webpage, undated.

9. Conclusion

In the research articulated in this report, we aimed to provide an updated overview of the commercial space industry, highlight recent changes in sectors most important to DoD, and describe new capabilities that are emerging and will grow in the coming years. Each chapter was devoted to a specific space mission and presented a market overview and an assessment of the key companies active in that sector. In this final chapter, we provide our overarching observations about the commercial space industry and briefly summarize each space mission.

Commercial Space Industry Observations

In this section, we discuss several overarching trends that we observe across the commercial space industry.

- **The more-established commercial space sectors are growing in capacity and capability**. The SATCOM sector has begun using HTS and is planning proliferated LEO constellations. The space launch sector has had two new NSSL-class entrants, is developing super-heavy LVs, and has a growing number of small LV entrants. The remote sensing sector has a quickly growing number of multi-satellite constellations, as well as diversity in sensor phenomenology and analytic products.
- New entrants are also responsible for recent growth in the commercial space industry. **The growth and evolution of new entrants has been driven by small satellite technologies and the proliferated constellation model, advanced manufacturing, use of AI/ML, and venture-capital investments**.
- Among the new space sectors, some will serve commercial space operators, while **other new space sectors will primarily target government customers**. Driven by the commercial proliferation of space, SDA entrants will offer enhanced collision warnings, and ground station entrants will provide data-transport services to offer timely downlink of high-volume data (i.e., for remote sensing satellites). Environmental monitoring entrants are collecting GNSS-RO data for NOAA and DoD. Space logistics entrants are planning space debris–removal services and on-orbit servicing for satellite life extension, both of which garner interest with government space programs.

Summaries by Space Mission

Each of the space sectors we reviewed in this report has unique characteristics driven by the history of its market and technological development. In the next sections, we summarize key findings from each commercial space sector. Table 9.1 follows with a list of the recent changes, issues, and futures to watch for in each commercial space sector.

SATCOM

The U.S. government has been a long-standing customer in the matured commercial SATCOM industry. However, the industry does not rely solely on the government. From 2014 to 2018, the revenue from the global government sector accounted for 21 percent of the total revenue for a set of major companies in the industry. Because of the increase in demand for data in general, commercial capacity is increasing at a substantial pace. The total available capacity from the major companies currently under contracts with DoD is estimated to be 254 Gbps, growing to 5,600 Gbps by 2022 (Bonds, Camm, and Willcox, forthcoming). Furthermore, the developments in the commercial NGSO constellations project will add 22,000 Gbps of global broadband capacity in the near future.

Space Launch

The NSSL-class launch market is mature under heavy government influence. Of the 233 NSSL-class launches conducted from 2006 to 2018, 176 were for U.S. government payloads, both military and civil. Therefore, the result of the upcoming NSSL Phase 2 contract award will shape the market.[1] The launch-service providers for this class of LVs are developing their capacity to support 48 to 60 launches per year, collectively. The market demand is estimated to be 17 to 29 per year, of which seven to nine are national security space launches. If the development of the new LVs are not delayed, commercial supply capacity will exceed DoD needs. However, there is a nontrivial probability of delays, which could cause a short-term supply shortage. Trends in launch technology, including reusability, on-orbit reignition, and increase in lift-capacity, may improve cost and launch responsiveness.

Remote Sensing

The commercial remote sensing market is evolving rapidly. Several established space-based remote sensing companies are leveraging improving technology to achieve new capabilities, such as finer spatial resolution. New entrants to the market are introducing proliferated constellations of dozens to several hundreds of satellites in LEO that will greatly expand the capacity beyond today's satellite-based Earth observation. Capabilities that are largely new to the commercial sector, such as SAR and space-based RF geolocation, are emerging. Commercial providers may use analytics, such as artificial intelligence, to improve the value proposition of Earth-observation data to the U.S. government and other potential customers. Although the government has had successful relationships with a limited number of commercial providers of satellite imagery, the changing landscape of commercial remote sensing creates additional opportunities to augment or free up capacity of DoD systems or fill capability gaps that DoD systems do not

---

[1] During the preparation of this report, we learned that the NSSL Phase 2 contract was awarded to ULA and SpaceX, and Northrop Grumman will discontinue development of the OmegA rocket (Erwin 2020a; Erwin 2020b).

meet. To take advantage of these additional opportunities, DoD will need to evaluate the operational risks associated with using commercial services. Although the global revenues for commercial remote sensing have approximately doubled over the past decade, the future size of the market is uncertain. Rapid change in the market can introduce risks in the long-term financial viability of individual companies or services.

Environmental Monitoring

Commercial capabilities in the environmental monitoring sector are made up of startup companies that rely heavily on government funding and thus focus on the government environmental monitoring mission. Among the most notable startups are those that have launched satellite constellations to collect GNSS-RO data that can inform terrestrial weather forecasts, climate monitoring, and space weather. However, the utility of these data is still under debate by DoD and NOAA. The open-source nature of environmental monitoring data is one shortfall in the business model for these startups, because the mission has historically been collaborative among many space institutions and international partnerships. If the commercial companies plan to sell their data to governments or other potential customers, this would present a marketing challenge to change from traditionally free open-source data. Although the commercial market has focused on GNSS-RO, which is a small area of environmental monitoring mission, other aspects of this sector have been neglected. Government interest in other types of environmental monitoring sensors, such as microwave or space weather sensors, may help spur new startups to emerge in other environmental monitoring market segments.

Space Domain Awareness

There are growing commercial opportunities to provide SDA services because of the growing number of companies and countries that have entered this market. The SDA mission can be broken into three main parts: data collection, data processing, and data sharing. Data collection and data processing are what we have focused our market research on, given DoD's existing capabilities for data sharing. It is unclear whether DoD is willing to outsource the data-processing portion of the SDA mission to commercial providers. If DoD is willing to outsource, several companies offer both data-collection and data-processing services. Establishing relationships with one or more of these companies may be an efficient and effective way for DoD to leverage the growing capabilities of the SDA commercial market. Finally, taking advantage of opportunities to increase capacity in SDA is critically important, because the solutions to capacity issues in other space missions trend toward proliferation.

Data Transmit/Receive Networks

Focusing on the GSaaS segment of the market, DoD has interest in leveraging commercial ground stations to augment the capacity of the AFSCN. Many commercial companies have built global ground station networks that would cater to government customers. The market extends beyond the government, because most GSaaS companies anticipate a growing demand from commercial satellite companies, especially remote sensing, with a growing need for large data downlinks. Many companies are exploring technological options, such as optical communications, phased-array antennas that can connect to multiple satellites, and connections to cloud computing resources, to improve existing antennas.

Space Logistics

The OoSS sector is a nascent market characterized by the development of on-orbit inspection, repair, replacement, refueling, orbital change, and upgrade capabilities. Currently, only a few companies exist in the global market, and many of these companies specialize in specific aspects of OoSS spacecraft, such as robotics and RPO. To date, only one company has launched an operational OoSS spacecraft, and that company is limited to providing life extension by rendezvous and orbit control. The OoSS capabilities are reportedly sought by GEO satellite operators to enhance fleet management through life extension. However, considerable technological development is still required for the full range of the OoSS capabilities to become operational and attract business from satellite operators.

Table 9.1. Recent and Future Developments in the Commercial Space Industry

Sector	Changes in Recent Years	Issues	Futures to Watch
Satellite communication	• Increased commercial capacity with increased market demand		• Added global broadband capacity from NGSO constellations
Space launch	• Increase in the number of launch service providers across all launch classes	• Possible commercial supply exceeding DoD needs • Possible launch delays causing launch supply shortage	• Technology developments: reusability, on-orbit reignition, increase lift capacity • Effect of NSSL Phase 2 contract award on market
Remote sensing	• Expansion in current and planned PLEO launches	• Uncertainty in commercial system's resilience in conflict environment	• Size of commercial market and financial viability of startups
Environmental monitoring	• NOAA and DoD focus on GNSS-RO • Success in some GNSS-RO launch and operations, commercial and government • Lack of progress in hyperspectral soundings	• Open-source nature of environmental monitoring data makes nongovernment customers unlikely • Uncertainty in utility of commercial GNSS-RO data	• New startups with developments in microwave, EO/IR, and space weather capabilities
SDA	• Increased demand and commercial supply with new market entrants	• Uncertainty in commercial system's resilience in conflict environment	• Size of commercial market and financial viability of startups • Space proliferation driving demand and/or collaboration
Data transmit/ receive network (ground stations)	• New U.S. companies offering GSaaS for commercial and government customers		• Electronically steered antennas/multiphase array • Optical communications technology
Space logistics (on-orbit servicing)	• Launch of only one company • Developing niche capabilities from a few companies	• Need more technological development for full range of OoSS capabilities	• Realization of technological developments, enabling on-orbit refueling, and assembly and manufacturing

SOURCE: RAND analysis of open-source reporting.

60

Recommendations

Considering both what we have observed in our overview of the commercial space industry and how a stakeholder, such as DoD, might use the information in this report, we provide the following recommendations:

- All space sectors we reviewed have experienced changes in the past five years (see Table 9.1), indicating that **it is important for DoD and other stakeholders to periodically update their information about the industry**.
- There are **several technology development and commercial viability factors that DoD and other stakeholders should track going forward** because these will significantly affect the space market (see Table 9.1).

Abbreviations

AFSCN	Air Force Satellite Control Network
AI/ML	artificial intelligence/machine learning
AWS	Amazon Web Services
CAMO	Commercial Augmented Mission Operations
DARPA	Defense Advanced Research Projects Agency
DoD	U.S. Department of Defense
EO/IR	electro-optical/infrared
FAA	Federal Aviation Administration
FCC	Federal Communications Commission
FSS	fixed satellite services
Gbps	gigabits per second
GEO	geosynchronous orbit
GNSS-RO	Global Navigation Satellite System Radio Occultation
GPS	Global Positioning System
GSaaS	ground station as a service
GSO	geosynchronous satellite operator
GTO	geostationary transfer orbit
HTS	high-throughput satellite
ISR	intelligence, surveillance, and reconnaissance
KSAT	Kongsberg Satellite Services
LEO	low earth orbit
LV	launch vehicle
MEO	medium earth orbit
MEV	Mission Extension Vehicle
NGA	National Geospatial-Intelligence Agency
NGSO	non–geosynchronous satellite operator
NOAA	National Oceanic and Atmospheric Agency
NRO	National Reconnaissance Office
NSS	National Security Space
NSSL	National Security Space Launch
OoSS	on-orbit satellite servicing
OSAM-1	On-orbit Servicing, Assembly, and Manufacturing 1
PLEO	proliferated low earth orbit
RF	radio frequency
RO	radio occultation

RPO	rendezvous and proximity operation
RSGS	Robotic Servicing of Geosynchronous Satellites
SAR	synthetic aperture radar
SATCOM	satellite communications
SBIR	Small Business Innovation Research
SDA	space domain awareness
SMC	Space and Missile Systems Center
SSA	space situational awareness
SSC	Swedish Space Corporation
SSO	sun-synchronous orbit
STM	space traffic management
SWIR	short-wave infrared
UDL	Unified Data Library
UHF	ultra high frequency
ULA	United Launch Alliance
USSF	U.S. Space Force
VHF	very high frequency

References

30th Space Wing, "Western Range Launch Database 2003–2018," database, September 2019, Not available to the general public.

45th Space Wing, "Eastern Range Launch Database 2003–2018," database, August 2019, Not available to the general public.

Air Force Research Laboratory, "Commercial Augmentation Service (CAS)," briefing, February 4, 2020, Not available to the general public.

Amazon Web Services, "AWS Space & Ground Station," briefing, May 27, 2020, Not available to the general public.

Astroscale, "ELSA-d," webpage, undated. As of October 22, 2021:
https://astroscale.com/elsa-d/

Atlas Space Operations, "Global Antenna Network," webpage, undated. As of June 10, 2020:
https://atlasground.com/antenna-network/

AWS—*See* Amazon Web Services.

Bhattacherjee, Debopam, Waqar Aqeel, Ilker Nadi Bozkurt, Anthony Aguirre, Balakrishnan Chandrasekaran, P. Brigten Godfrey, Gregory Laughlin, Bruce Maggs, and Ankit Singla, "Gearing Up for the 21st Century Space Race," *HotNets '18: Proceedings of the 17th ACM Workshop on Hot Topics in Networks*, Redmond, Wash., November 15–16, 2018. As of August 20, 2021:
https://users.cs.duke.edu/~bmm/assets/pubs/BhattacherjeeABACGLMS18.pdf

BlackSky, "Products and Services," webpage, undated. As of May 1, 2020:
https://blacksky.com/products-services/

Blue Origin, "New Glenn," webpage, undated. As of March 1, 2020:
https://www.blueorigin.com/new-glenn/

Bonds, Timothy A., Frank Camm, and Jordan Willcox, *Ensuring Theater Satellite Communications: Capabilities and Costs of Commercial Services*, Santa Monica, Calif.: RAND Corporation, RR-A103-1, forthcoming.

Carey, Mike, "Why the 'Amazonification' of Satellite Data Communications Is a Good Thing," *SpaceNews*, December 11, 2018. As of August 23, 2021:
https://spacenews.com/why-the-amazonification-of-satellite-data-communications-is-a-good-thing/

Clark, Colin, "Exclusive: War in Space 'Not a Fight Anybody Wins'—Gen. Raymond," *Breaking Defense*, April 6, 2017. As of June 11, 2020:
https://breakingdefense.com/2017/04/exclusive-war-in-space-not-a-fight-anybody-wins-gen-raymond/

Code of Federal Regulations 47, Section 25.161, Automatic Termination of Station Authorization, October 1, 2014. As of August 30, 2021:
https://www.law.cornell.edu/cfr/text/47/25.161

Cohen, Rachel S., "Space Traffic Transition to Commerce Hits Speed Bumps," *Air Force Magazine*, February 14, 2020. As of June 11, 2020:
https://www.airforcemag.com/space-traffic-transition-to-commerce-hits-speed-bumps/

Crunchbase, "GeoOptics," webpage, undated. As of June 11, 2020:
https://www.crunchbase.com/organization/geooptics

Davenport, Justin, "NGIS OmegA Fires for Two Minutes in First Static Test – Nozzle Incident Under Review," *NASA Space Flight*, May 30, 2019. As of March 1, 2020:
https://www.nasaspaceflight.com/2019/05/ngis-omega-fires-first-test-nozzle-incident-review/

Davis, Joshua P., John P. Mayberry, and Jay P. Penn, *On-Orbit Servicing: Inspection, Repair, Refuel, Upgrade and Assembly of Satellites in Space*, Arlington, Va.: Center for Space Policy and Strategy, April 2019. As of November 20, 2021:
https://aerospace.org/sites/default/files/2019-05/Davis-Mayberry-Penn_OOS_04242019.pdf

Eenmaa, Sven, "Investment Perspectives: Conferring on On-Orbit Satellite Servicing," *ISS 360: The ISS National Lab Blog*, November 15, 2018. As of March 13, 2010:
https://www.issnationallab.org/blog/investment-perspectives-conferring-on-on-orbit-satellite-servicing/

Erwin, Sandra, "In-Orbit Services Poised to Become Big Business," *SpaceNews*, June 10, 2018. As of May 5, 2020:
https://spacenews.com/in-orbit-services-poised-to-become-big-business/

Erwin, Sandra, "ULA, SpaceX, Blue Origin, Northrop Grumman Submit Bids for National Security Launch Procurement Contract," *SpaceNews*, August 12, 2019a. As of March 1, 2020:
https://spacenews.com/ula-spacex-blue-origin-northrop-grumman-submit-bids-for-national-security-launch-procurement-contract/

Erwin, Sandra, "Air Force: SSA Is No More; It's 'Space Domain Awareness,'" *SpaceNews*, November 14, 2019b. As of June 11, 2020:
https://spacenews.com/air-force-ssa-is-no-more-its-space-domain-awareness/

Erwin, Sandra, "Northrop Grumman Touts Financial Strength in Marketing Pitch for OmegA Rocket," *SpaceNews*, December 3, 2019c. As of March 1, 2020: https://spacenews.com/northrop-grumman-touts-financial-strength-in-marketing-pitch-for-omega-rocket/

Erwin, Sandra, "Starlink's Busy Launch Schedule Is Workable, Says 45th Space Wing," *SpaceNews*, January 7, 2020a. As of February 6, 2020: https://spacenews.com/starlinks-busy-launch-schedule-is-workable-says-45th-space-wing/

Erwin, Sandra, "DARPA Picks Northrop Grumman as Its Commercial Partner for Satellite Servicing Program," *SpaceNews*, March 4, 2020b. As of March 13, 2020: https://spacenews.com/darpa-picks-northrop-grumman-as-its-commercial-partner-for-satellite-servicing-program/

Erwin, Sandra, "Space Fence Surveillance Radar Site Declared Operational," *SpaceNews*, March 28, 2020c. As of June 11, 2020: https://spacenews.com/space-fence-surveillance-radar-site-declared-operational/

Erwin, Sandra, "Pentagon Picks SpaceX and ULA to Remain Its Primary Launch Providers," *SpaceNews*, August 7, 2020d. As of December 8, 2020: https://spacenews.com/pentagon-picks-spacex-and-ula-to-launch-national-security-satellites-for-next-five-years/

Erwin, Sandra, "Northrop Grumman to Terminate OmegA Rocket Program," *SpaceNews*, September 9, 2020e. As of December 9, 2020b: https://spacenews.com/northrop-grumman-to-terminate-omega-rocket-program/

Espinosa, Shellie-Anne, "National Space Defense Center Transitions to 24/7 Operations," Air Force Space Command (Archived) webpage, January 26, 2018. As of August 23, 2021: https://www.afspc.af.mil/News/Article-Display/Article/1423932/national-space-defense-center-transitions-to-247-operations/

Etherington, Darrell, "Watch OneWeb Launch 34 Satellites for Its Broadband Constellation Live," *Tech Crunch*, February 6, 2020. As of February 6, 2020: https://techcrunch.com/2020/02/06/watch-oneweb-launch-34-satellites-for-its-broadband-constellation-live/

Euroconsult, "First Industry Report on Earth Observation Data Distribution Trends and Strategies," press release, Paris, Montreal, and Washington, D.C., March 27, 2014. As of August 23, 2021: https://www.euroconsult-ec.com/press-release/first-industry-report-on-earth-observation-data-distribution-trends-and-strategies/

Euroconsult, *Satellite-Based Earth Observation Market Prospects to 2028*, 12th ed., Paris, 2019.

Euroconsult, "Euroconsult Research Projects Smallsat Market to Nearly Quadruple over Next Decade," press release, Paris, Washington, D.C., Montreal, Yokohama, August 5, 2019. As of August 31, 2021:
https://www.euroconsult-ec.com/press-release/euroconsult-research-projects-smallsat-market-to-nearly-quadruple-over-next-decade/

Eutelsat, "Satellites," webpage, undated. As of February 21, 2020:
https://www.eutelsat.com/en/satellites.html

Eutelsat, "Successful Launch of Eutelsat Konnect," press release, Paris, January 17, 2020. As of August 23, 2021:
https://www.businesswire.com/news/home/20200116005928/en/Successful-Launch-Of-EUTELSAT-KONNECT

Eutelsat Communications Group, *Consolidated Financial Statements as of 30 June 2019*, Paris, undated. As of November 11, 2019:
https://www.eutelsat.com/files/PDF/investors/EC_comptes%20consolid%c3%a9s_annuels_FY19_EN.pdf

FAA Commercial Space Transportation and the Commercial Space Transportation Advisory Committee, *2015 Commercial Space Transportation Forecasts*, Washington, D.C.: Federal Aviation Administration, Office of Commercial Space Transportation, April 2015. As of August 23, 2021:
https://brycetech.com/reports/report-documents/Commercial_Space_Transportation_Forecasts_2015.pdf

FCC—*See* Federal Communications Commission.

Federal Aviation Administration, "Licensed Launches," webpage, August 18, 2020. As of August 28, 2020:
http://www.faa.gov/data_research/commercial_space_data/launches/

Federal Communications Commission, "Order and Declaratory Ruling: In the Matter of WorldVu Satellites Limited, Petition for a Declaratory Ruling Granting Access to the U.S. Market for the OneWeb NGSO FSS System," Washington, D.C., FCC-17-77, June 22, 2017a. As of August 23, 2021:
https://docs.fcc.gov/public/attachments/FCC-17-77A1_Rcd.pdf

Federal Communications Commission, "Report and Order and Further Notice of Proposed Rulemaking: In the Matter of Update to Parts 2 and 25 Concerning Non-Geostationary, Fixed-Satellite Service Systems and Related Matters," Washington, D.C., FCC-17-122, September 26, 2017b. As of August 23, 2021:
https://docs.fcc.gov/public/attachments/FCC-17-122A1.pdf

Federal Communications Commission, "Order and Declaratory Ruling: In the Matter of Telesat Canada, Petition for Declaratory Ruling to Grant Access to the U.S. Market for Telesat's NGSO Constellation," Washington, D.C., FCC-17-147, November 2, 2017c. As of August 23, 2021:
https://docs.fcc.gov/public/attachments/FCC-17-147A1.pdf

Federal Communications Commission, "FCC Application for Space and Earth Station: MOD or AMD," Washington, D.C., File Number SAT−MOD−20180319−00022, March 19, 2018a. As of August 23, 2021:
https://licensing.fcc.gov/myibfs/download.do?attachment_key=1357209

Federal Communications Commission, "Memorandum Opinion, Order and Authorization: In the Matter of Space Exploration Holdings, LCC, Application for Approval for Orbital Deployment and Operating Authority for the SpaceX NGSO Satellite System," Washington, D.C., FCC-18-38, March 28, 2018b. As of August 23, 2021:
https://docs.fcc.gov/public/attachments/FCC-18-38A1.pdf

Federal Communications Commission, "Memorandum Opinion, Order and Authorization in the Matter of Space Exploration Holdings, LCC, Application for Approval for Orbital Deployment and Operating Authority for the SpaceX V-Band NGSO Satellite System," Washington, D.C., FCC-18-161, November 15, 2018c. As of August 23, 2021:
https://docs.fcc.gov/public/attachments/FCC-18-161A1.pdf

Federal Communications Commission, "Order and Authorization in the Matter of Space Exploration Holdings, LCC, Request for Modification of the Authorization for the SpaceX NGSO Satellite System," Washington, D.C., FCC-DA-19-342, April 26, 2019a. As of August 23, 2021:
https://docs.fcc.gov/public/attachments/DA-19-342A1.pdf

Federal Communications Commission, "Application for Modification, in the Matter of WorldVu Satellites Limited, Modification to OneWeb U.S. Market Access Grant for the OneWeb Ku- and Ka-Band System," Washington, D.C., SAT-MPL-20200526-00062, May 26, 2020. As of August 23, 2021:
https://fcc.report/IBFS/SAT-MPL-20200526-00062/2379569.pdf

Ferster, Warren, "DigitalGlobe Closes GeoEye Acquisition," *SpaceNews*, January 31, 2013. As of August 23, 2021:
https://spacenews.com/digitalglobe-closes-geoeye-acquisition/

Firefly Aerospace, "Firefly Alpha," webpage, undated. As of March 5, 2020:
https://firefly.com/launch-alpha/

Forrester, Chris, "Telesat LEO Constellation in 2022," *Advanced Television*, March 2, 2020. As of March 10, 2020:
https://advanced-television.com/2020/03/02/telesat-leo-constellation-in-2022/

Foust, Jeff, "May the Satellite Industry Live in Interesting Times," *Space Review*, September 17, 2018. As of May 21, 2020:
https://www.thespacereview.com/article/3570/1

Foust, Jeff, "Rethinking Satellite Servicing," *The Space Review*, February 4, 2019a. As of May 25, 2020:
https://www.thespacereview.com/article/3653/1

Foust, Jeff, "Small Launch Vehicle Companies See Rideshares as an Opportunity and a Threat," *SpaceNews*, February 7, 2019b. As of November 18, 2020:
https://spacenews.com/small-launch-vehicle-companies-see-rideshare-as-an-opportunity-and-a-threat/

Foust, Jeff, "Relativity Space Raises $140 million," *SpaceNews*, October 1, 2019c. As of March 5, 2020:
https://spacenews.com/relativity-space-raises-140-million/

Foust, Jeff, "Acting NOAA Leader Stresses Importance of Public-Private Partnerships," *SpaceNews*, January 22, 2020a. As of August 23, 2021:
https://spacenews.com/acting-noaa-leader-stresses-importance-of-public-private-partnerships

Foust, Jeff, "Virgin Orbit Nearing First Launch," *SpaceNews*, February 5, 2020b. As of March 5, 2020:
https://spacenews.com/virgin-orbit-nearing-first-launch/

Foust, Jeff, "Opportunities Grow for Smallsat Rideshare Launches," *SpaceNews*, February 6, 2020c. As of March 12, 2020:
https://spacenews.com/opportunities-grow-for-smallsat-rideshare-launches/

GeoOptics, "GeoOptics Celebrates Two Years of On-Orbit Operations," press release, January 15, 2020. As of June 11, 2020:
https://geooptics.com/geooptics-celebrates-two-years-of-on-orbit-operations/

Griggs, Mary Beth, "Two Commercial Satellites Just Docked in Space for the First Time," *The Verge*, February 26, 2020. As of March 9, 2020:
https://www.theverge.com/2020/2/26/21154426/commercial-satellites-docking-space-northrop-grumman-intelsat

Halsband, Arie, "Pioneering Last-Mile Logistics in Space," *Effective Space Solutions*, February 2017. As of August 23, 2021:
https://25ce73ad-d841-4d59-a48c-7f2ebabac6e1.filesusr.com/ugd/58bc6d_d6a1cc93baaf402eb47837fbc27376dc.pdf

Harebottle, Adrienne, "Heading into the LEO Revolution," *Via Satellite*, February 2020a. As of February 6, 2020:
http://interactive.satellitetoday.com/via/february-2020/heading-into-the-leo-revolution/

Harebottle, Adrienne, "The New, Holistic View of Space," *Via Satellite*, March 2020b. As of May 21, 2020:
http://interactive.satellitetoday.com/via/march-2020/the-new-holistic-view-of-space/

Henry, Caleb, "Commercial Laser Comm Edges Closer to Reality," *SpaceNews*, June 26, 2018. As of August 23, 2021:
https://spacenews.com/commercial-laser-comm-edges-closer-to-reality/

Henry, Caleb, "Lockheed Martin Mulls Electronically Steered Antennas for Verge Ground Station Expansion," *SpaceNews*, May 8, 2019a. As of August 23, 2021:
https://spacenews.com/lockheed-martin-mulls-electronically-steered-antennas-for-verge-ground-station-expansion/

Henry, Caleb, "Musk Says Starlink 'Economically Viable' with Around 1,000 Satellites," *SpaceNews*, May 15, 2019b. As of May 6, 2020:
https://spacenews.com/musk-says-starlink-economically-viable-with-around-1000-satellites/

Henry, Caleb, "Fresh $18.7 Million Funding Round Puts PlanetiQ Weather Constellation Back on Track," *SpaceNews*, July 11, 2019c. As of August 23, 2021:
https://spacenews.com/fresh-18-7-million-funding-round-puts-planetiq-weather-constellation-back-on-track

Henry, Caleb, "Dankberg Teases ViaSat-4 Specs, Still Mulling MEO Constellation," *SpaceNews*, October 16, 2019d. As of February 21, 2020:
https://spacenews.com/dankberg-teases-viasat-4-specs-still-mulling-meo-constellation/

Henry, Caleb, "ULA Gets Vague on Vulcan Upgrade Timeline," *SpaceNews*, November 20, 2019e. As of March 1, 2020:
https://spacenews.com/ula-gets-vague-on-vulcan-upgrade-timeline/

Henry, Caleb, "Maxar Wins $142 Million NASA Robotics Mission," *SpaceNews*, January 31, 2020a. As of March 13, 2020:
https://spacenews.com/maxar-wins-142-million-nasa-robotics-mission/

Henry, Caleb, "Geostationary Satellite Orders Bouncing Back," *SpaceNews*, February 21, 2020b. As of May 21, 2020:
https://spacenews.com/geostationary-satellite-orders-bouncing-back/

Henry, Caleb, "Airbus Impressed by Northrop Grumman, but Remains Undecided on Satellite Servicing," *SpaceNews*, March 11, 2020c. As of May 25, 2020:
https://spacenews.com/airbus-impressed-by-northrop-grumman-but-remains-undecided-on-satellite-servicing/

Hitchens, Theresa, "Crider: SSA Data 'Library' Will Open to Allies," *Breaking Defense*, May 3, 2019a. As of June 11, 2020:
https://breakingdefense.com/2019/05/crider-ssa-data-library-will-open-to-allies/

Hitchens, Theresa, "Hill Nixes Trump Space Tracking Plan," *Breaking Defense*, December 19, 2019b. As of June 11, 2020:
https://breakingdefense.com/2019/12/hill-nixes-trump-space-tracking-plan/

Hitchens, Theresa, "SpaceLogistics Sat Servicing Mission Taps New Markets," *Breaking Defense*, March 20, 2020. As of May 25, 2020:
https://breakingdefense.com/2020/03/spacelogistics-sat-servicing-mission-taps-new-markets/

Holmes, Mark, "Satellite Servicing Becomes an Actual Market," *Via Satellite*, March 2019a. As of March 9, 2020:
http://interactive.satellitetoday.com/via/march-2019/satellite-servicing-becomes-an-actual-market/

Holmes, Mark, "Satellite Manufacturers Enter a Whole New World," *Via Satellite*, September 19, 2019b. As of May 25, 2020:
https://www.satellitetoday.com/business/2019/09/19/satellite-manufacturers-enter-a-whole-new-world/

Howell, Elizabeth, "Firefly Aerospace Preps for Debut Flight of Its Alpha Rocket in April," *Space*, January 6, 2020. As of March 5, 2020:
https://www.space.com/firefly-aerospace-alpha-rocket-first-launch-april-2020.html

Inmarsat, "Fleet Data," webpage, undated-a. As of February 21, 2020:
https://www.inmarsat.com/service/fleet-data

Inmarsat, "Global Xpress," webpage, undated-b. As of March 20, 2020:
https://www.inmarsat.com/service/global-xpress/

Inmarsat, *In Touch: Inmartsat PLC Annual Report and Accounts 2014*, London, March 2015. As of November 11, 2019:
https://www.inmarsat.com/content/dam/inmarsat/corporate/documents/result-centre/Inmarsat_ARA-2014.pdf.coredownload.pdf

Inmarsat, *Building Momentum: Inmarsat PLC Annual Report and Accounts 2016*, London, March 2017. As of November 11, 2019:
https://www.inmarsat.com/wp-content/uploads/2019/12/2016-Annual-Report.pdf

Inmarsat, *Enabling Connectivity: Annual Report and Accounts 2018*, London, March 2019. As of November 11, 2019:
https://www.inmarsat.com/content/dam/inmarsat/corporate/documents/result-centre/Inmarsat_Annual_Report_2018.pdf.coredownload.pdf

Intelsat, "Intelsat Coverage Maps," webpage, undated-c. As of February 21, 2020:
http://www.intelsat.com/global-network/satellites/fleet/

Intelsat, *Intelsat Epic^{NG}*, September 2016a. As of November 25, 2021:
https://www.intelsatgeneral.com/wp-content/uploads/2016/09/6493-Epic-Positioning_2018.pdf

Intelsat, *Annual Report Pursuant to Section 13 or 15(d) of the Securities Exchange Act of 1934, for the Fiscal Year Ended December 31, 2015*, Washington, D.C.: U.S. Securities and Exchange Commission, commission file number 001-35878, May 2, 2016b. As of November 11, 2019:
http://media.corporate-ir.net/media_files/IROL/13/131114/intelsat_performance_summary_2015/pdf/Intelsat2015AnnualShareholderLetter_Form20-F.pdf

Intelsat, *Annual Report Pursuant to Section 13 or 15(d) of the Securities Exchange Act of 1934, for the Fiscal Year Ended December 31, 2017*, Washington, D.C.: U.S. Securities and Exchange Commission, commission file number 001-35878, February 2018. As of November 11, 2019:
https://investors.intelsat.com/static-files/87d4727d-d09e-4f62-9b92-a726f69a58fd

Intelsat, *Annual Report Pursuant to Section 13 or 15(d) of the Securities Exchange Act of 1934 for the Fiscal Year Ended December 31, 2018*, Washington, D.C.: U.S. Securities and Exchange Commission, commission file number 001-35878, February 20, 2019. As of November 11, 2019:
https://investors.intelsat.com/static-files/6fce8ad2-6c78-4974-8940-5de84aafd421

Iridium Communications, *2018 Annual Report*, McLean, Va., 2019. As of November 11, 2019:
https://investor.iridium.com/download/2018+Annual+Report+%28Final+EZBlue%29.pdf

Jewett, Rachel, "U.S. Bankruptcy Court Approves OneWeb Sale to UK Government, Bharti," *Via Satellite*, October 5, 2020. As of December 8, 2020:
https://www.satellitetoday.com/business/2020/10/05/u-s-bankruptcy-court-approves-oneweb-sale-to-uk-government-bharti/

Joint Publication 3-14, *Space Operations*, Washington, D.C.: Joint Chiefs of Staff, April 10, 2018, Incorporating Change 1, October 26, 2020. As of August 20, 2021:
https://www.jcs.mil/Portals/36/Documents/Doctrine/pubs/jp3_14ch1.pdf?ver=qmkgYPyKBvsIZyrnswSMCg%3D%3D

Kasaboski, Dallas, "Satellite Social Distancing," Northern Sky Research website, March 31, 2020. As of June 23, 2020:
https://www.nsr.com/ssa-satellite-social-distancing/

Kim, Yool, Mary Lee, George Nacouzi, Brian Dolan, Moon Kim, and Thomas Light, *A Framework for an Integrated Assessment of Commercial Space Capabilities*, Santa Monica, Calif.: RAND Corporation, forthcoming, Not available to the general public.

Kim, Yool, Ellen Pint, David Galvan, Meagan Smith, Therese Marie Jones, and William Shelton, *How Can DoD Better Leverage Commercial Capabilities? Understanding Business Processes and Practices in the Commercial Satellite Service Industry*, Santa Monica, Calif.: RAND Corporation, 2016, Not available to the general public.

Kongsberg Satellite Services, "Ground Network Services," webpage, undated. As of August 23, 2021:
https://www.ksat.no/services/ground-station-services/

KSAT—*See* Kongsberg Satellite Services.

Lal, Bhavya, Asha Balakrishnan, Becaja Caldwell, Reina Buenconsejo, and Sara Carioscia, *Global Trends in Space Situational Awareness (SSA) and Space Traffic Management (STM)*, Washington, D.C.: IDA Science and Technology Policy Institute, IDA Document D-9074, April 2018. As of May 28, 2020:
https://www.ida.org/-/media/feature/publications/g/gl/global-trends-in-space-situational-awareness-ssa-and-space-traffic-management-stm/d-9074.ashx

LeoLabs, "Global Phased-Array Radar Network," webpage, undated. As of August 31, 2021:
https://www.leolabs.space/radars/

Leone, Dan, "NOAA Told to Consider Commercial Data in New U.S. Space Weather Strategy," *SpaceNews*, November 4, 2015. As of August 23, 2021:
https://spacenews.com/noaa-told-to-consider-commercial-data-in-new-u-s-space-weather-strategy/

Mann, Adam, "Rocket Lab's Electron Rocket," *Space*, October 3, 2019. As of March 5, 2020:
https://www.space.com/electron-rocket.html

Maxar, "OSAM-1 and Spider," webpage, undated. As of May 25, 2020:
https://explorespace.maxar.com/moon/osam-1-and-spider/

McLeod, Gary, Ellen Pint, Eric Larson, Eder Sousa, and Jonathan Tran, *U.S. Space Launch Locations to Support the National Security Space Launch Program: An Independent Assessment of the Ability of the Eastern and Western Ranges to Support Forecasted Launch Demands*, Santa Monica, Calif.: RAND Corporation, RR-4272-AF, 2019, Not available to the general public.

NASA—*See* National Aeronautics and Space Administration.

National Aeronautics and Space Administration, "What Is Remote Sensing," NASA Earth Data, April 20, 2020. As of June 9, 2020:
https://earthdata.nasa.gov/learn/remote-sensing

NewSpace Index, "Small Satellite Launchers," webpage, undated-a. As of August 31, 2021:
https://www.newspace.im/launchers

NewSpace Index, "Welcome to NewSpace Index," website, undated-b. As of March 1, 2020:
https://www.newspace.im/

Northrop Grumman, "SpaceLogistics," webpage, undated. As of October 21, 2021:
https://www.northropgrumman.com/space/space-logistics-services/

Nyirady, Annamarie, "HawkEye 360 Provides RF Data to BlackSky," *Via Satellite*, June 3, 2019. As of August 23, 2021:
https://www.satellitetoday.com/imagery-and-sensing/2019/06/03/hawkeye-360-provides-rf-data-to-blacksky/

Office of the President, *National Space Policy of the United States of America*, Washington, D.C., June 28, 2010. As of November 15, 2020:
https://obamawhitehouse.archives.gov/sites/default/files/national_space_policy_6-28-10.pdf

Office of the President, *National Security Strategy of the United States of America*, Washington, D.C., December 2017. As of November 15, 2020:
https://www.whitehouse.gov/wp-content/uploads/2017/12/NSS-Final-12-18-2017-0905.pdf

Office of Space Commerce, "ClearView Arrangements Awarded to Three Remote Sensing Firms," press release, March 29, 2003.

OneWeb, "Technology," webpage, undated. As of February 23, 2020:
https://www.oneweb.world/technology

OneWeb Satellites, "Revolutionizing the Economics of Space," webpage, undated. As of February 6, 2020:
https://onewebsatellites.com/

OrbitFab, "Products & Hardware," webpage, undated. As of March 9, 2020:
https://www.orbitfab.space/products

Pereira, John J., Robert Atlas, Joanne Ostroy, William J. Blackwell, Thomas S. Pagano, Jacob Inskeep, and Mark Seymour, "NOAA's CubeSat-Related Activities for Gap Mitigation and Future Planning," briefing, 31st Annual Small Satellite Conference, Logan, Utah, August 8, 2017. As of August 23, 2021:
https://digitalcommons.usu.edu/cgi/viewcontent.cgi?article=3637&context=smallsat

Ponnappan, Raman, "Additive Manufacturing in Launch Vehicles," *Spacetech Asia*, August 30, 2018. As of November 18, 2020:
https://www.spacetechasia.com/additive-manufacturing-in-launch-vehicles/

Public Law 87-624, Communications Satellite Act of 1962, August 31, 1962.

Public Law 102-555, Land Remote Sensing Policy Act of 1992, October 28, 1992.

Puteaux, Maxime, "Satellite Manufacturing Faces Changes, Uncertainty in Coming Years," *Via Satellite*, May 20, 2020. As of May 21, 2020:
https://www.satellitetoday.com/launch/2020/05/20/satellite-manufacturing-faces-changes-uncertainty-in-coming-years/

Puteaux, Maxime, and Alexandre Najjar, "Analysis | Are Smallsats Entering the Maturity Stage?" *SpaceNews*, August 6, 2019. As of August 31, 2021:
https://spacenews.com/analysis-are-smallsats-entering-the-maturity-stage/

RBC Signals, "Locations," webpage, undated. As of June 10, 2020:
http://rbcsignals.com/#section_locations

Relativity Space, "Terran 1," webpage, undated. As of March 5, 2020:
https://www.relativityspace.com/terran

Rocket Lab, "Completed Missions," webpage, undated. As of March 5, 2020:
https://www.rocketlabusa.com/missions/completed-missions/

Salmi, Bryce, "The World's Largest 3D Metal Printer Is Churning Out Rockets," *IEEE Spectrum*, October 25, 2019. As of March 5, 2020:
https://spectrum.ieee.org/aerospace/space-flight/the-worlds-largest-3d-metal-printer-is-churning-out-rockets

Satellite Industry Association, *State of the Satellite Industry Report*, Washington, D.C., 2010–2019. As of August 23, 2021:
https://sia.org/news-resources/state-of-the-satellite-industry-report/

Satellite Industry Association, *2019 State of the Satellite Industry Report*, executive summary, Washington, D.C., May 2019. As of August 23, 2021:
https://brycetech.com/reports/report-documents/SSIR-2019-2-pager.pdf

SBIR STTR, "Tethers Unlimited, Inc.," Small Business Innovation Research webpage, undated. As of November 18, 2020:
https://www.sbir.gov/sbc/tethers-unlimited-inc

de Selding, Peter B., "NGA Letters Cast Cloud over GeoEye's EnhancedView Funding," *SpaceNews*, June 23, 2012. As of August 23, 2021:
https://spacenews.com/nga-letters-cast-cloud-over-geoeyes-enhancedview-funding/

SES, "Explore the Full Fleet," webpage, undated. As of February 21, 2020:
https://www.ses.com/our-coverage#/explore

SES, *FY 2014 Results: Year Ended 31 December 2014*, February 20, 2015. As of November 11, 2019:
https://www.ses.com/sites/default/files/2016-11/150219_FY_2014_ANALYST_PRESENTATION_FINAL_0.pdf

SES, *Full Year 2016 Results: Year Ended 31 December 2016*, London, February 24, 2017. As of November 11, 2019:
https://www.ses.com/sites/default/files/2017-02/170223_FY2016_Analyst%20Presentation_FINAL.pdf

SES, *Consolidated Financial Statements as at and for the Year Ended 31 December 2018 and Independent Auditor's Report*, Betzdorf, Luxembourg, March 1, 2019. As of November 11, 2019:
https://www.ses.com/sites/default/files/2019-03/190301_Financial%20statements%20FY%202018%20and%20audit%20report.pdf

Spacelink website, undated. As of February 23, 2020:
https://www.starlink.com/

SpaceNews staff, "DigitalGlobe Loses WorldView-4 to Gyro Failure," *SpaceNews*, January 7, 2019. As of August 23, 2021:
https://spacenews.com/digitalglobe-loses-worldview-4-satellite-to-gyro-failure/

SpaceX, "Falcon 9," webpage, undated-a. As of March 1, 2020:
https://www.spacex.com/vehicles/falcon-9/

SpaceX, "Falcon Heavy," webpage, undated-b. As of March 1, 2020:
https://www.spacex.com/vehicles/falcon-heavy/

Spire Global, "Spire Taps AWS Ground Station to Extend Ground Station Network," press release, Las Vegas and San Francisco, November 27, 2018. As of June 11, 2020:
https://www.globenewswire.com/news-release/2018/11/27/1657796/0/en/Spire-Taps-AWS-Ground-Station-to-Extend-Ground-Station-Network.html

SSC—*See* Swedish Space Corporation.

Stoffler, Ralph, "United States Air Force Space Weather," briefing, Washington, D.C.: Headquarters, U.S. Air Force, May 2, 2017. As of August 23, 2021: https://www.swpc.noaa.gov/sites/default/files/images/u33/SWW%202017%20DoD%20Space%20Weather%20-%20Final.pdf

Strout, Nathan, "How the NRO Learned to Stop Worrying and Love the Commercial Imagery," *Air Force Times*, June 4, 2019a. As of August 23, 2021: https://www.airforcetimes.com/c2-comms/satellites/2019/06/04/how-the-nro-learned-to-stop-worrying-and-love-the-commercial-imagery/

Strout, Nathan, "The Pentagon Wants to Extend the Life of Satellites and Refuel on Orbit," C4ISRNet, October 1, 2019b. As of May 23, 2020: https://www.c4isrnet.com/battlefield-tech/space/2019/10/01/the-pentagon-wants-to-extend-the-life-of-satellites-and-refuel-on-orbit/

Strout, Nathan, "The Pentagon Wants a Roadside Assistance Service in Space," C4ISRNet, February 5, 2020. As of May 23, 2020: https://www.c4isrnet.com/battlefield-tech/space/2020/02/05/the-pentagon-wants-a-roadside-assistance-service-in-space/

Swedish Space Corporation, "SSC's Global Ground Station Network," webpage, undated. As of June 10, 2020: https://www.sscspace.com/ssc-worldwide/ground-station-network/

Telesat, *Annual Report Pursuant to Section 13 or 15(d) of the Securities Exchange Act of 1934 for the Fiscal Year Ended December 31, 2018*, Washington, D.C.: U.S. Securities and Exchange Commission, commission file number 333-159793-01, March 1, 2019. As of November 11, 2019: https://www.sec.gov/Archives/edgar/data/1465191/000161577419003425/s115704_20f.htm

Telesat, "Telesat LEO – Why LEO?" webpage, undated. As of February 23, 2020: https://www.telesat.com/services/leo/why-leo

Tethers Unlimited, "In-Space Services," webpage, undated. As of March 9, 2020: https://www.tethers.com/in-space-services/

Triezenberg, Bonnie L., Colby Peyton Steiner, Grant Johnson, Jonathan Cham, Eder Sousa, Moon Kim, and Mary Kate Adgie, *Assessing the Impact of U.S. Air Force National Security Space Launch Acquisition Decisions: An Independent Analysis of the Global Heavy Lift Launch Market*, Santa Monica, Calif.: RAND Corporation, RR-4251-AF, April 2020. As of August 23, 2021: https://www.rand.org/pubs/research_reports/RR4251.html

ULA—*See* United Launch Alliance.

United Launch Alliance, "About," webpage, undated. As of March 1, 2020:
https://www.ulalaunch.com/about

United Launch Alliance, "Rocket Rundown: A Fleet Overview," technical summary, 2019. As of March 1, 2020:
https://www.ulalaunch.com/docs/default-source/rockets/atlas-v-and-delta-iv-technical-summary.pdf

U.S. Government Accountability Office, "Defense: DoD Commercial Satellite Communication Procurements," in *2016 Annual Report: Additional Opportunities to Reduce Fragmentation, Overlap, and Duplication and Achieve Other Financial Benefits*, Washington, D.C., GAO-16-375, April 13, 2016, pp. 44–46. As of August 23, 2021:
https://www.gao.gov/assets/gao-16-375sp.pdf

Van Ryswyk, Martin, "RapidEye Constellation to Be Retired in 2020," Planet, January 16, 2020. As of August 23, 2021:
https://www.planet.com/pulse/rapideye-constellation-to-be-retired-in-2020/

Viasat, "KA-SAT Satellite," webpage, undated-a. As of February 5, 2020:
https://www.viasat.com/products/high-capacity-satellites

Viasat, "Satellite Fleet," webpage, undated-b. As of October 22, 2021:
https://www.viasat.com/space-innovation/satellite-fleet/

Viasat, "ViaSat Announces Record $1.1 Billion in Revenues and $1.4 Billion in Awards for Fiscal 2013," press release, Carlsbad, Calif., May 16, 2013. As of May 16, 2020:
https://www.viasat.com/news/record-11-billion-revenues-and-14-billion-awards-for-fiscal-2013

Viasat, *Annual Report 2015*, Carlsbad, Calif., September 2015. As of November 11, 2019:
http://investors.viasat.com/static-files/5bf60371-4407-4b96-a45e-096ae8cfc708

Viasat, *Annual Report 2017*, Carlsbad, Calif., September 2017. As of November 11, 2019:
http://investors.viasat.com/static-files/29e4d5fe-7624-4ca0-a03b-1b323778fe4a

Viasat, *Annual Report 2019*, Carlsbad, Calif., September 2019. As of November 11, 2019:
http://investors.viasat.com/static-files/743e5c27-c611-4a1c-ab86-63b66b82451b

Viasat, "Real-Time Earth: Rethinking Ground Segment as a Service," brochure, Carlsbad, Calif., 2020. As of June 10, 2020:
https://www.viasat.com/sites/default/files/media/documents/viasat-real-time-earth-brochure.pdf

Virgin Orbit, "LauncherOne," webpage, undated. As of June 4, 2020:
https://virginorbit.com/technology/

Werner, Debra, "Lofty Aspiration for Spire's Weather-Watching Cubesats," *SpaceNews*, September 17, 2015. As of August 23, 2021:
https://spacenews.com/lofty-aspirations-for-spires-weather-watching-cubesats/

Werner, Debra, "Government Agencies Prepare for Piggyback Flights, Secondary Payloads," *SpaceNews*, September 17, 2018. As of March 5, 2020:
https://spacenews.com/government-agencies-prepare-for-piggyback-flights-secondary-payloads/

Werner, Debra, "Are Laser Links Ready for Prime Time?" *SpaceNews*, May 22, 2019. As of August 23, 2021:
https://spacenews.com/optical-communications-space-tech-expo/

Werner, Debra, "NOAA Signals Strong Appetite for Radio Occultation," *SpaceNews*, January 15, 2020. As of August 23, 2021:
https://spacenews.com/radio-occultation-ams-2020/

White House, "Fact Sheet: U.S. Commercial Remote Sensing Policy," press release, Washington, D.C.: U.S. Commercial Remote Sensing Space Policy, May 13, 2003. As of August 23, 2021:
https://georgewbush-whitehouse.archives.gov/news/releases/2003/05/20030513-8.html

White House, "Space Policy Directive-3, National Space Traffic Management Policy," presidential memorandum, June 18, 2018. As of June 11, 2020:
https://www.whitehouse.gov/presidential-actions/space-policy-directive-3-national-space-traffic-management-policy/

CPSIA information can be obtained
at www.ICGtesting.com
Printed in the USA
LVHW060120050622
720468LV00013B/312